Hayk Sedrakyan
Nairi Sedrakyan

Competition math for middle school: must-knows and beyond

2023

About the authors

Hayk Sedrakyan is an IMO medal winner, professional mathematical Olympiad coach in greater Boston area, Massachusetts, USA. He has been the dean (and one of the main developers) of one of the biggest math competition preparation programs in the USA. He has been a Professor of mathematics in Paris and has a PhD in mathematics (optimal control and game theory) from the UPMC - Sorbonne University, Paris, France. Hayk is a Doctor of mathematical sciences in USA, France, Armenia and holds three master's degrees in mathematics from institutions in Germany, Austria, Armenia and has spent a small part of his PhD studies in Italy. Hayk Sedrakyan has worked as a scientific researcher for the European Commission (sadco project), has been one of the Team Leaders at Harvard-MIT Mathematics Tournament (HMMT) and was an invited speaker at Imperial College London. He took part in the International Mathematical Olympiads (IMO) in United Kingdom, Japan and Greece. Hayk has been elected as the President of the students' general assembly and a member of the management board of the *Cite Internationale Universitaire de Paris* (10,000 students, 162 different nationalities) and the same year they were nominated for the Nobel Peace Prize. Hayk Sedrakyan is the son and was one of the students of Nairi Sedrakyan.

Nairi Sedrakyan is involved in national and international Mathematics Olympiads having been the President of Armenian Mathematics Olympiads and a member of the IMO problem selection committee. Nairi Sedrakyan was the winner of **Erdös Award** 2022 (one of the highest international awards in mathematics) for contributing to the development of mathematics worldwide, in average 1 person per year (globally) wins this award. He is one of 3 people from entire ex-USSR territory ever to win Erdös Award, the other 2 are: Grigori Perelman's professor (the only person in the world who coached 2 Fields Medal winners) and Nikolay Konstantinov (founder of Tournament of the Towns). Nairi Sedrakyan is the author of **the most difficult problem ever proposed in the history of the International Mathematical Olympiad (IMO)**, 5th problem of 37th IMO. This problem is considered to be the hardest problems ever in IMO because none of the members of the strongest teams (national Olympic teams of China, USA, Russia) succeeded to solve it correctly and because national Olympic team of China (the strongest team in IMO) obtained a cumulative result equal to 0 points and was ranked 6th in the final ranking of the countries instead of the usual 1st or 2nd place. The British 2014 film X+Y, released in the USA as *A Brilliant Young Mind*, inspired by the film *Beautiful Young Minds* (focuses on an English mathematical genius chosen to represent the United Kingdom at IMO) also states that this problem is the hardest problem ever proposed in the history of IMO (minutes 9:40-10:30). His students have received 20 medals in International Mathematical Olympiad, including Gold and Silver medals. Nairi Sedrakyan received a Gold Medal for contributions to World's Mathematical Olympiads and Scientific Activities.

Any comments or suggestions?
Then, please contact **sedrakyan.hayk@gmail.com**

Overview

This book includes **author-created math competition problems with author-prepared solutions (never published before)**. The book also includes theory. The main goal of the book is to teach problem solving strategies, how to solve non-standard problems and how to score better on middle school math competitions. It is intended as a teacher's manual of mathematics, as a manual for math competition coaches, a self-study handbook for middle-school students and mathematical competitors.

Keywords: Competition math for middle school: must knows and beyond, American Mathematics Competitions preparation.

Disclaimer: A few of these problems were proposed in different national competitions (including Armenia). The names of the authors of these few problems (and the exact years) is unknown to us for providing better reference, as there is no online library or reference preserved. So, we listed them in the section "problems from national competitions". All the information is provided and cited to the best of our knowledge.

Any comments or suggestions?
Then, please contact **sedrakyan.hayk@gmail.com**

Mathematical competition is not about winning or losing, it is about mastering the art of thinking creatively and smart.

Hayk Sedrakyan.

Contents

1	**Parity (theory)**	**10**
	1.1 Practice problems	13
	1.1.1 Solutions of practice problems	15
	1.2 Problems from national competitions	21
	1.2.1 Hints and Solutions	22
2	**Divisibility and remainders**	**23**
	2.1 Practice problems	28
	2.1.1 Solutions of practice problems	31
	2.2 Problems from national competitions	47
	2.2.1 Hints and Solutions	50
3	**Pigeonhole principle**	**56**
	3.1 Practice problems	58
	3.1.1 Solutions of practice problems	60
	3.2 Problems from national competitions	68
	3.2.1 Hints and Solutions	69
4	**Combinatorics**	**71**
	4.1 Practice problems	74
	4.1.1 Solutions of practice problems	77
	4.2 Problems from national competitions	87
	4.2.1 Hints and Solutions	90
5	**Counting in two different ways**	**95**
	5.1 Practice problems	97
	5.1.1 Solutions of practice problems	98

6 Triangle inequality — 103
- 6.1 Practice problems — 104
  - 6.1.1 Solutions of practice problems — 106
- 6.2 Problems from national competitions — 119
  - 6.2.1 Hints and solutions — 121

7 Boundary principle — 125
- 7.1 Practice problems — 126
  - 7.1.1 Solutions of practice problems — 128

8 Graphs — 134
- 8.1 Practice problems — 141
  - 8.1.1 Solutions of practice problems — 143

9 Solving integer equations — 151
- 9.1 Practice problems — 155
  - 9.1.1 Solutions of practice problems — 156

10 Invariants — 161
- 10.1 Practice Problems — 163
  - 10.1.1 Solutions of Practice Problems — 164

11 Mathematics in games — 169
- 11.1 Practice Problems — 173
  - 11.1.1 Solutions of Practice Problems — 175

12 Proving inequalities — 182
- 12.1 Practice Problems — 184
  - 12.1.1 Proofs — 185
- 12.2 Practice Problems — 188
  - 12.2.1 Proofs — 189
- 12.3 Practice Problems — 195
  - 12.3.1 Proofs — 196

13 Geometry — 202
- 13.1 Practice Problems — 204
  - 13.1.1 Solutions of Practice Problems — 205
- 13.2 Practice Problems — 216

- 13.2.1 Solutions of Practice Problems 217
- 13.3 Practice Problems 230
  - 13.3.1 Solutions of Practice Problems 231

Acknowledgment

The author would like to thank his family for the support.

To Margarita, Nairi, Ani, Jane and Luna.

The author would like to thank *Zhenya Sukiasyan* for the cover and for the illustrations.

Chapter 1

Parity (theory)

Recall that an integer is called **even** if it is a double of another integer.
Example of some even numbers. 2, 4, 2022, -10, 0.

An integer is called **odd** if it is not a double of another integer.
Example of some odd numbers. 1, 3, 2023, -1, -3.

The fact (of a number) being *even* or *odd* is called **parity** of a number. So

- Any even number can be written as $2 \cdot k$, where k is an integer.
- Any odd number can be written as $2 \cdot k + 1$, where k is an integer.

In this chapter we consider examples demonstrating how the **concept of parity** can be used to solve non-standard problems.

Example 1.1:

There are 1, 3 or 5 peach fruits on each of 20 peach trees of a garden. Can the total number of all peach fruits of this garden be equal to 61?

Example 1.1: Solution
Answer. No
Solution: The total number of all peach fruits of this garden cannot be equal to 61, as on each peach tree there is an odd number of fruits and the sum of 20 odd numbers is an even number. So, it cannot be equal to 61.

Example 1.2:

A cube is on a table. A child turns the cube around one of its four edges that touch the table. Can the child, after 25 turns, bring this cube to its initial location on the table (even if in a rotated position)?

Example 1.2: Solution

Answer. No

Solution: Imagine that the table is colored as a chessboard, such that each square of the chessboard is of the same size as a face of the cube (see the picture).

Let in the beginning the cube was placed on any white square of the table. Then, after one turn the cube is placed on a black square of the table. This means that after each turn the color of the square (that the cube is placed on) is changing (from white to black or from black to white). Thus, after 25 turns, the cube will be placed on a black square.

So, after 25 turns, the child cannot bring it to its initial position.

Example 1.3:

Two opposite corner squares of the same diagonal of a chessboard were removed. Is it possible to cover the remaining board by 31 dominos (figures composed of two squares)?

Example 1.3: Solution

Answer. No

Solution: Let the remaining board (62 squares) be covered by 31 dominos. Note that each domino covers one black and one white square of the board. Then, 31 dominos divide 62 squares of the board to 31 pairs of squares, so that 31 of them are white and the other 31 are black. Given that two removed squares have the same color. So, the number of white squares is 30 or 32. This is not possible, as there are 31 white squares.

So, it is not possible to cover the remaining board by 31 dominos.

1.1 Practice problems

Problem 1.1. *There are eight houses on one side of a road. We know that the number of people living in two neighboring houses differs by one. Is it possible that the overall number of people living in all houses is 29?*

Problem 1.2. *Is it possible to change some of $\star$ symbols to $+$ symbols and the rest of them to $-$ symbols so that this equation becomes correct?*

$$1 \star 2 \star \ldots \star 30 = 0.$$

Problem 1.3. *There are 51 towns in a kingdom. The king ordered to build roads so that there are five incoming roads for each town. Can they do it?*

Problem 1.4. *Is it possible that all 11 wheels (see the picture) spin at the same time?*

Problem 1.5. *Prove that the sum of any 30 different numbers from $1, 2, \ldots, 58$ is not equal to the sum of the other 28 numbers from $1, 2, \ldots, 58$.*

Problem 1.6. *Is it possible to divide a convex heptagon (seven sided shape) into parallelograms?*

Problem 1.7. *Is it possible to write a positive integer in each square of 3×4 rectangle so that the sum of all numbers in each row is prime and the sum of all numbers in each column is prime?*

Problem 1.8. *Is it possible to move a chess knight from the bottom left corner of the chessboard to the upper right corner by passing through each square of the chessboard only once?*

Problem 1.9. *Prove that it is not possible to cover any 10×10 chessboard with 25 such ⌐⌐ figures.*

Problem 1.10. *Is it possible to color the vertices of a 2022 − gon (a polygon with 2022 sides) in red or in white so that the number of sides with endpoints of the same color is equal to the number of sides with endpoints of different colors?*

Problem 1.11. *There are three hockey pucks (disks) on the skating rink (skating area), whose centers are not on the same line. Andrew hits one of the pucks such that its center is on the other side of the line passing through the centers of two other pucks. After 25 hits, is it possible to return every puck to its initial position?*

1.1.1 Solutions of practice problems

Problem 1.1: *There are eight houses on one side of a road. We know that the number of people living in two neighboring houses differs by one. Is it possible that the overall number of people living in all houses is 29?*

> Answer. It is not possible.
> **Solution**: Let the houses be numbered (in order) by numbers $1, \ldots, 8$. Given that the number of people living in the houses 1 and 2 is odd. Similarly, for each pair of 3 and 4, 5 and 6, 7 and 8 the number of people living in both houses is odd. So, the number of people living in all eight houses is the same as the sum of four odd numbers. The sum of four odd numbers is always even, so it cannot be equal to 29.

Problem 1.2: *Is it possible to change some of $\star$ symbols to $+$ symbols and the rest of them to $-$ symbols so that this equation becomes correct?*

$$1 \star 2 \star \ldots \star 30 = 0.$$

> Answer. It is not possible.
> **Solution**: Note that the *sum* and the *difference* of any two integers have the same parity. So, the parity of $1 \star 2 \star \ldots \star 30$ is always the same. Also $1 + 2 + \ldots + 30 = (1 + 30) + (2 + 29) + \ldots + (15 + 16) = 31 \cdot 15$ is an odd number. So, $1 \star 2 \star \ldots \star 30$ is an odd number and cannot be equal to 0.

Problem 1.5: *There are 51 towns in a kingdom. The king ordered to build roads so that there are five incoming roads for each town. Can they do it?*

> Answer. It is not possible.
> **Solution**: If there are 5 incoming roads in each town, such that they do not pass through another town, then the number of all roads is $\frac{51 \cdot 5}{2}$. This is not possible as $\frac{51 \cdot 5}{2}$ is not an integer.

Problem 1.4: *Is it possible that all 11 wheels (see the picture) spin at the same time?*

Answer. It is not possible.

Solution: Let the first wheel spins in the positive direction (see the picture). Then, the second wheel spins in the negative direction, the third wheel spins in the positive direction, and so on. So, 11th wheel spins in the positive direction.

We have that wheels 1 and 11 spin (at the same time) in the positive direction (which is not possible). The case when wheel 1 spins in the negative direction is also not possible.

Problem 1.5: *Prove that the sum of any 30 different numbers from $1, 2, \ldots, 58$ is not equal to the sum of the other 28 numbers from $1, 2, \ldots, 58$.*

Proof: Note that $1+2+\ldots+58 = (1+58)+(2+57)+\ldots+(29+30) = 59 \cdot 29$ is an odd number. Let S be the sum of 30 different numbers from these 58 numbers so that this sum is equal to the sum of the other 28 numbers, then $1 + 2 + \ldots + 58 = 2 \cdot S$. As $2 \cdot S$ is an even number, so it cannot be equal to an odd number.

Problem 1.6: *Is it possible to divide a convex heptagon (seven sided shape) into parallelograms?*

Answer. It is not possible.
Solution: Let a convex heptagon be divided into parallelograms. Let us choose some side AB from the heptagon (see the picture).

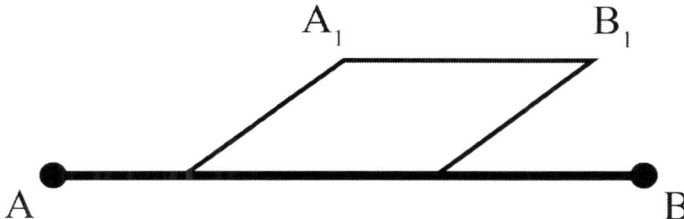

Then, there must be a parallelogram such that one of its sides is on AB. Let us denote by A_1B_1 one of the sides of the parallelogram that is parallel to AB. Similarly, we have that there is a parallelogram such that one of its sides is on A_1B_1, if A_1B_1 is not on one of the sides of the heptagon (see the picture).

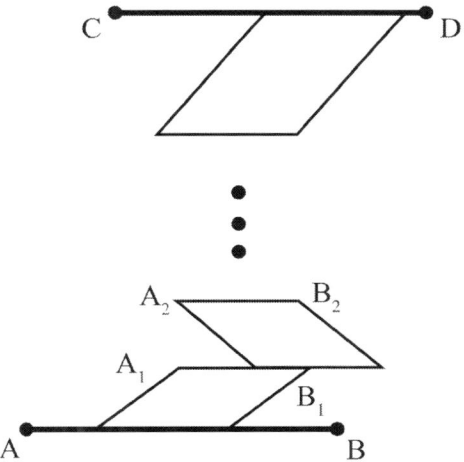

Let A_2B_2 be the side of the second parallelogram that is parallel to A_1B_1. Continuing like this we get that there is a parallelogram such that one of its sides is parallel to AB and belongs to side CD of the heptagon. So, we got that the sides of this heptagon should be pairwise parallel to each other, which is not possible as heptagon has an odd number of sides.

Problem 1.7: *Is it possible to write a positive integer in each square of 3×4 rectangle so that the sum of all numbers in each row is prime and the sum of all numbers in each column is prime?*

> Answer. It is not possible.
> **Solution**: Let the sum of the numbers in the ith row be p_i, where $i = 1, 2, 3$, and the sum of the numbers in j^{th} column be q_j, where $j = 1, 2, 3, 4$.
> From one side, the sum of all numbers in the table is $S = p_1 + p_2 + p_3$, where $p_j \geq 4$, $i = 1, 2, 3$. Note that p_1, p_2, p_3 are odd numbers (odd primes), so $S = p_1 + p_2 + p_3$ is an odd number.
> From the other side, $S = q_1 + q_2 + q_3 + q_4$, where $q_j \geq 3$, $j = 1, 2, 3, 4$. Note that q_1, q_2, q_3, q_4 are odd numbers (odd primes). So, S is an even number. This is not possible, as we got that S must be an odd number.

Problem 1.8: *Is it possible to move a chess knight from the bottom left corner of the chessboard to the upper right corner by passing through each square of the chessboard only once?*

> Answer. It is not possible.
> **Solution**: The knight must do 63 moves to pass through every square of the chessboard. Note that the initial square (where the knight is) is black. Note also that after each move the color of the square (where the knight is) is changing from black to white or from white to black. So, after 63rd move the knight is on a white square. On the other hand, the upper right square of the chessboard is black.
> So, it is not possible to move a chess knight from the bottom left corner of the chessboard to the upper right corner by passing through each square of the chessboard only once.

Problem 1.9: *Prove that it is not possible to cover any 10×10 chessboard by 25 such ⌐⊥⌐ figures.*

> **Proof**: Let us color the square as a chessboard (see the picture).
>
>
>
> Note that each such figure can cover four squares of the 10×10 board. Moreover, the number of white squares of the figure is 1 or 3.
>
> If it is possible to cover 10×10 board by 25 such figures, then we have an odd number of white squares (as the sum of 25 odd numbers is odd). However, the number of white squares on a chess board is 50. This ends the proof.

Problem 1.10: *Is it possible to color the vertices of a $2022 - gon$ (a polygon with 2022 sides) in red or in white so that the number of sides with endpoints of the same color is equal to the number of sides with endpoints of different colors?*

> Answer. It is not possible.
> **Solution**: Assume there is a coloring satisfying the conditions of the problem. Let us erase the vertices that have the same color. Assume that the remaining sides of the polygon are consecutively connected. So, we have a $1011 - gon$ polygon such that any of its sides has endpoints of different colors. This is not possible, as if we number the endpoints of this polygon by $1, 2, \ldots, 1011$, then the endpoints 1 and 1011 have the same color. On the other hand, they are the endpoints of the same side, so they must have different colors. This is not possible.

Problem 1.11: *There are three hockey pucks (disks) on the skating rink (skating area), whose centers are not on the same line. Andrew hits one of the pucks such that its center is on the other side of the line passing through the centers of two other pucks. After 25 hits, is it possible to return every puck to its initial position?*

Answer. It is not possible.

Solution: Let the centers of pucks be A, B, C. According to the condition of the problem A, B, C do not belong to the same line. After each hit two of the points stay in their previous position, while the third changes its position such that A, B, C are not on the same line.

Let us look at the triangle ABC after each step. If before hitting we read the letters A, B, C in the negative direction, then after the hit we read them in the positive direction.

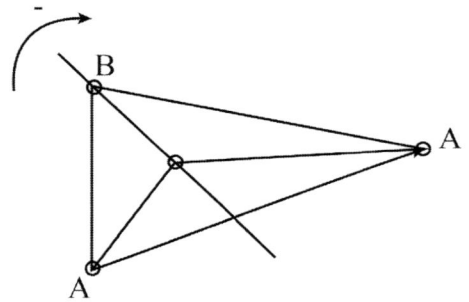

So, if in the beginning we read the letters A, B, C in the negative direction, then after the 25th hit we read them in the positive direction. So, after 25 such hits, the pucks cannot return to their initial position.

1.2 Problems from National Competitions

Problem 1: *Find the number of natural numbers n, such that $n^2 + 15n + 9$ is divisible by 18.*

Problem 2: *Given that a is not an integer and $a+b, b+c, a+c$ are integers. Can $ab + bc + ac$ be an integer?*

Problem 3: *Assume that the sum of two numbers a and b is odd. Let the set of natural numbers be divided into two sets: set A and set B. Prove that there are two numbers in any of these two sets such that their difference is equal to a or b. (N.M. Sedrakyan)*

Problem 4: *Find the sum of numbers $p, q, p^2 + q^3, p^3 + q^2$ if they are all prime numbers. (N.M. Sedrakyan)*

Problem 5: *Assume p and q are both natural numbers and q is odd. Prove that there do not exist natural numbers a and b, where $a \neq b$, such that $\frac{1}{a} + \frac{1}{b} + \frac{1}{a-b} = \frac{p}{q}$. (N.M. Sedrakyan)*

1.2.1 Hints and Solutions

Problem 1: *Find the number of positive integers n, such that $n^2 + 15n + 9$ is divisible by 18.*

> Answer: 0
> **Hint**: Prove that $(n(n+15)+9)$ is an odd number for any integer n.

Problem 2: *Given that a is not an integer and $a+b, b+c, a+c$ are integers. Can $ab + bc + ac$ be an integer?*

> **Hint**: $ab + bc + ac = (a+b)(a+c) - a^2$.

Problem 3: *Assume that the sum of two numbers a and b is odd. Let the set of natural numbers be divided into two sets: set A and set B. Prove that there are two numbers in any of these two sets such that their difference is equal to a or b. (N.M. Sedrakyan)*

> **Hint**: Proof by **contradiction argument** (assume that what we want to prove is not true and then get consequences that are not correct). Assume $a \in A$, $2a \in B$, $3a \in A$,..., $b \cdot a \in A$. If $b \in A$, then $2b \in B$,..., $ab \in B$, which is not correct. So $b \in B$. Then, consider the number $a + b$.

Problem 4: *Find the sum of numbers $p, q, p^2 + q^3, p^3 + q^2$ if they are all prime numbers. (N.M. Sedrakyan)*

> Answer. 53.
> **Hint**: Note that one of the numbers p or q must be even.

Problem 5: *Assume p and q are both natural numbers and q is odd. Prove that there do not exist natural numbers a and b, where $a \neq b$, such that $\frac{1}{a} + \frac{1}{b} + \frac{1}{a-b} = \frac{p}{q}$. (N.M. Sedrakyan)*

> **Hint**: Proof by contradiction argument.
> Let a and b be chosen such that the common denominator of $\frac{1}{a} + \frac{1}{b} + \frac{1}{a-b}$ is an odd number. Show that a and b are even numbers.

Chapter 2
Divisibility and remainders

Integer a is called **_divisible_** by an integer b (where $b \neq 0$), if there exists an integer c so that $a = b \cdot c$.

b is called a **_divisor_** (**_factor_**) of a and a is called a **_multiple_** of b.

The fact that integer a is divisible by b is written as $a \vdots b$.
Examples of divisibility. $15 \vdots 3$ and $2022 \vdots 6$.

A number is called **prime**, if it has exactly two different positive factors.
Examples of prime numbers. $2,3,5,7,11,13,17,19$ are prime numbers.

A number is called **composite**, if it has more than two positive factors.
Examples of composite numbers. $4,6,8,12,15$ are composite numbers.

Remark. 1 has one divisor, so it is neither prime nor composite.

Theorem 2.1 (Fundamental theorem of Arithmetic). *Any integer greater than 1 is possible to represent in exactly one way as a product of prime numbers (apart from rearrangement of the order of these primes).*

> **Example 2.1:**
>
> *Find the sum of all prime divisors of 2022.*

> **Example 2.1: Solution**
> Answer. 342
> **Solution:** We have that $2022 = 2 \cdot 3 \cdot 337$. Numbers $2, 3, 337$ are prime and according to the Fundamental theorem of Arithmetic this representation is unique. So we have $2 + 3 + 337 = 342$.

> **Example 2.2:**
>
> *Let a and b be natural numbers, so that $2a = 17b$. Prove that $a - b$ is a composite number.*

> **Example 2.2: Proof**
> **Proof:** We have that $17b$ is even, so b is also even. Let $b = 2k$, where k is a natural number. From $2a = 17b$ we have that $a = 17k$. So $a - b = 17k - 2k = 15k$, where $(a-b) \vdots 1, (a-b) \vdots 3, (a-b) \vdots 5$. So $a - b$ is composite.

The following divisibility properties are used to solve a large number of problems. Here, letters represent natural numbers.

1. If $a \vdots b$ and $b \vdots c$, then $a \vdots c$.

2. If $a \vdots b$, then $ac \vdots b$.

3. If $a \vdots b$, then $ac \vdots bc$.

4. If $a \vdots b$, then $a \geq b$.

5. If $a \vdots b$ and $b \vdots a$, then $a = b$.

6. If $a \vdots c$ and $b \vdots c$ $(a > b)$, then $(a + b) \vdots c$ and $(a - b) \vdots c$.

7. If p is prime and $ab \vdots p$, then $a \vdots p$ or $b \vdots p$.

8. Let $d_1, d_2, \ldots, d_k$ be the divisors of n and $1 = d_1 < d_2 < \ldots < d_k = n$.

9. d_2 is prime

10. For any $i \in \{1, 2, \ldots, k\}$ we have $d_i \cdot d_{k+1-i} = n$.

    The greatest of all common divisors of two integers a and b is called the **greatest common divisor** (gcd) of these two numbers.
    It is usually denoted as (a, b) or $\gcd(a, b)$.

    If the *greatest common divisor* of a and b is 1, then these numbers are called **relatively prime**.

> **Example 2.3:**
>
> *At most how many composite numbers is possible to choose from $100, 101, \ldots, 500$, so that any two of the chosen numbers are relatively prime?*

Example 2.3: Solution
Answer. 8
Solution: We have that $23 \cdot 23 = 529 > 500$, so if $n \leq 500$ is a composite number, then according to the property 9 we have $d_2 < 23$. We have that any chosen composite number is divisible by any of numbers $2, 3, 5, 7, 11, 13, 17, 19$. From the statement above it follows that the number of composite numbers that meet the requirement is not more than 8. Some examples of such numbers are: $121, 125, 128, 169, 217, 243, 289, 361$

The smallest (least) of all common multiples of numbers a and b is called **least common multiple** of a and b.
It is usually denoted as $[a, b]$ or $\mathrm{lcm}(a, b)$.

11. If $a > b$ then $(a, b) = (a - b, b)$.

12. $(a,b) \cdot [a,b] = ab$.

13. There exist *unique* integers q, r so that $a = b \cdot q + r$, where $0 \leq r < b$.

    Then, we say that after dividing a by b we get a **quotient** q and a **remainder** r.

    > **Example 2.4:**
    > *2022 is divided by each of the numbers $1, 2, \ldots, 1000$. Find the greatest of all remainders.*

    > **Example 2.4: Solution**
    > **Answer.** 672
    > **Solution:** Let 2022 be divided by a number a, where $1 \leq a \leq 1000$. The quotient is q and the greatest remainder is r.
    > We have that $2022 = a \cdot q + r$, $a \geq r + 1$ and $q \geq 2$.
    > So $2022 \geq (r+1)q + r \geq 2(r+1) + r$, then $r \leq 673$.
    > When $r = 673$, $2022 = a \cdot q + 673$. So $1349 = aq$, So $a \in \{1, 19, 71\}$, which is not possible. Thus $r \leq 672$.
    > Note that $2022 = 675 \cdot 2 + 672$.

14. If $b \vdots a$ and $c \vdots a$, then $(b, c) \vdots a$.

15. If $a \vdots b$ and $a \vdots c$, then $a \vdots [b, c]$.

Example 2.5:

Find the number of all pairs of positive integers (m, n), so that $(12m - 1) \vdots n$ and $(12n - 1) \vdots m$.

Example 5: Solution
Answer. 7

Solution: If $m = n$, then we have $(12m - 1) \vdots m$. So $m = 1$. So, the pair $(1, 1)$ is a solution.

Note that if (m, n) satisfies the assumption of the problem, then (n, m) also satisfies the assumption of the problem.

It is sufficient to find all pairs (m, n) so that $m < n$. Thus $(12m - 1) \vdots n$ and $12m - 1 < 12n - 1$. This means that $12m - 1 = n$ or $12m - 1 = 5n$ or $12m - 1 = 7n$ or $12m - 1 = 11n$.

If $n = 12m - 1$, then from $(12n - 1) \vdots m$ we get $(144m - 13) \vdots m$. So $m = 1$ and $m = 13$ and the solutions are $(1, 11), (13, 155)$.

If $5n = 12m - 1$, then $(60n - 5) \vdots m$. So $17 \vdots m$, this is not possible.

If $7n = 12m - 1$, then we get that it is not possible.

If $12m - 1 = 11n$, then we get the pair $(23, 25)$.

So, there are seven pairs satisfying the conditions of the problem.

2.1 Practice problems

Problem 2.1. *Find the number of factors (divisors) of the numbers:*

(a) 2022.

(b) 1200.

(c) 1000000.

Problem 2.2. *Prove that the sum of two consecutive natural numbers is odd and their product is even.*

Problem 2.3. *Prove that the product of three consecutive numbers is divisible by 6.*

Problem 2.4. *Find the smallest natural number n so that the product of the first n numbers is divisible by 420.*

Problem 2.5. *How many zeros does the product $1 \cdot 2 \cdot \ldots \cdot 100$ end with?*

Problem 2.6. *Two of the digits of a natural number are ones and all others are zeros. Can that number be a perfect square (square of a natural number)?*

Problem 2.7. *Prove that $x^2 - y^2 = 2022$ has no natural number solution.*

Problem 2.8. *Prove that $x^3 - y^3 = 2022$ has no natural number solution.*

Problem 2.9. *What is the greatest number of consecutive natural numbers, so that their sum is equal to 2022?*

Problem 2.10. *A natural number is called "interesting" if it is possible to write a new number divisible by 7 by only using some of its digits, but not all of them. Find the greatest number that is not "interesting". (N.M. Sedrakyan)*

Problem 2.11. *Find the number of natural numbers n so that $2^n + 5^n - 65$ is a perfect square. (N.M. Sedrakyan)*

Problem 2.12. *Numbers $1, 2, \ldots, 9$ are written in a 3×3 board, so that in every square there is one number. Then the sum of each row, column and diagonal was calculated. How many of the calculated sums can be prime?*

Problem 2.13. *Find the product of natural numbers x, y, z, t so that $68(xyzt + xy + zt + xt + 1) = 157(yzt + y + t)$.*

Problem 2.14. *Find the product of prime numbers p, q and r so that $p(p-5) + q(q-5) = r(r+5)$.*

Problem 2.15. *Find the smallest possible value of n so that both n and $n + 2015$ are squares of a natural number. (N.M. Sedrakyan)*

Problem 2.16. *The sum of five natural numbers is 595. What is the greatest number of zeros that their product can end with? (N.M. Sedrakyan)*

Problem 2.17. *Find the smallest natural number that is divisible by 11 and the sum of its digits is 19.*

Problem 2.18. *Integers a, b, c are greater than 1 and have the following properties: $(a+b) \vdots (c+1)$, $(b+c) \vdots (a+1)$ and $(a+c) \vdots (b+1)$. Find the greatest possible value of $a + b + c$.*

Problem 2.19. *For numbers a, b we have: $(a, b) + [a, b] = a + b + 1874$. Find the smallest possible value of $a + b$.*

Problem 2.20. *Suppose m and n are natural numbers so that $(m+n, 10) = 1$. Prove that $\frac{m}{n}$ is possible to represent in the form*

$$\frac{10^{k-1}a_1 + 10^{k-2}a_2 + \ldots + a_k}{10^{k-1}b_1 + 10^{k-2}b_2 + \ldots + b_k},$$

where k is a natural number, $a_i, b_i \in \{1, 2, \ldots, 9\}$ and $a_i + b_i = 9$ for $i = 1, \ldots, k$.

Problem 2.21. *A natural number is written in each square of 8×8 board. All numbers are pairwise different. It is known that any two numbers written in two neighboring squares are not relatively prime. Find the smallest possible value of the greatest of these numbers.*

Problem 2.22. *Find the number of pairs (p,q), where p and q are primes, so that $p^4 - q$ and $q^4 - p$ are also prime.*

2.1.1 Solutions of practice problems

Problem 2.1: *Find the number of factors (divisors) of the numbers:*

(a) 2022.

(b) 1200.

(c) 1000000.

Solution:

Answer *(a)*: 8

We have $2022 = 2 \cdot 3 \cdot 337$. So, the divisors are:
$1, 2, 3, 337, 2 \cdot 3, 3 \cdot 337, 2 \cdot 337, 2 \cdot 3 \cdot 337$.

Answer *(b)*: 30

We have $1200 = 2^4 \cdot 3 \cdot 5^2$. So each divisor of 1200 is of the form $2^a \cdot 3^b \cdot 5^c$, where $a \in \{0, 1, 2, 3, 4\}, b \in \{0, 1\}, c \in \{0, 1, 2\}$. According to the Fundamental Theorem of Arithmetic, the number of integers of the form $2^a \cdot 3^b \cdot 5^c$ is the same as the number of triplets (a, b, c). The number of such triplets is $5 \cdot 2 \cdot 3 = 30$.

Answer *(c)*: 49

We have $1000000 = 2^6 \cdot 5^6$. Thus, each divisor of 1000000 is of the form $2^a 5^b$, where $a, b \in \{0, 1, 2, 3, 4, 5, 6\}$. So, the number of divisors of 1000000 is $7 \cdot 7 = 49$.

Problem 2.2: *Prove that the sum of two consecutive natural numbers is odd and their product is even.*

Proof: Let us consider two consecutive numbers: n and $n + 1$.
We have $n + (n + 1) = 2n + 1$ is odd.
Note that one of the numbers n or $n + 1$ is even, then according to Property 2 we have $n(n + 1)$ is also even.

Problem 2.3: *Prove that the product of three consecutive numbers is divisible by 6.*

> **Proof**: According to property 2 and problem 2, the product of three consecutive numbers is even.
> Note that if $k \vdots 3$, then according to property 6 we have $(k+3) \vdots 3$. Additionally, there are two whole numbers between k and $k+3$, namely $k+1$ and $k+2$. So the product of any three consecutive numbers is divisible by 3.
> According to the property 2 their product is also divisible by 3.
> According to property 15, the product of any three consecutive numbers is divisible by $[2,3] = 6$.

Problem 2.4: *Find the smallest natural number n so that the product of the first n numbers is divisible by 420.*

> Answer: 7
> **Solution**:
> We have $420 = 2^2 \cdot 3 \cdot 5 \cdot 7$. As $(1 \cdot 2 \cdot \ldots \cdot n) \vdots 420$, then according to property 1, $1 \cdot 2 \cdot \ldots \cdot n \vdots 7$. From the last statement, using property 7, we get that one of the numbers from $1, 2, \ldots, n$ is divisible by 7. Applying property 4 we get $n \geq 7$.
> From the other hand $1 \cdot 2 \cdot 3 \cdot 4 \cdot 5 \cdot 6 \cdot 7 = (420 \cdot 12) \vdots 420$

Problem 2.5: *How many zeros does the product* $1 \cdot 2 \cdot \ldots \cdot 100$ *end with?*

Answer: 24
Solution:
We have $1 \cdot 2 \cdot \ldots \cdot 100 =$
$= 5 \cdot 10 \cdot 15 \cdot \ldots \cdot 100 \cdot (1 \cdot 2 \cdot 3 \cdot 4) \cdot (6 \cdot 7 \cdot 8 \cdot 9) \cdot \ldots \cdot (96 \cdot 97 \cdot 98 \cdot 99) =$
$= 5^{20} \cdot 1 \cdot 2 \cdot \ldots \cdot 20 \cdot (1 \cdot 2 \cdot 3 \cdot 4) \cdot (6 \cdot 7 \cdot 8 \cdot 9) \cdot \ldots \cdot (96 \cdot 97 \cdot 98 \cdot 99) =$
$= 5^{20} \cdot 5 \cdot 10 \cdot 15 \cdot 20 \cdot (1 \cdot 2 \cdot 3 \cdot 4)^2 \cdot (6 \cdot 7 \cdot 8 \cdot 9)^2 \cdot (11 \cdot 12 \cdot 13 \cdot 14)^2 \cdot$
$\cdot (16 \cdot 17 \cdot 18 \cdot 19)^2 \cdot (21 \cdot 22 \cdot 23 \cdot 24) \cdot \ldots \cdot (96 \cdot 97 \cdot 98 \cdot 99) =$
$= 5^{24} \cdot (1 \cdot 2 \cdot 3 \cdot 4)^3 \cdot (6 \cdot 7 \cdot 8 \cdot 9)^2 \cdot \ldots \cdot (96 \cdot 97 \cdot 98 \cdot 99) =$
$= 5^{24} \cdot 2^{24} \cdot a$, where a is a natural number and $(a, 5) = 1$.
So, the number $1 \cdot 2 \cdot 3 \cdot \ldots \cdot 100$ ends with 24 zeros.

Problem 2.6: *Two of the digits of a natural number are ones and all others are zeros. Can that number be a perfect square (square of a natural number)?*

Answer: It is impossible
Solution:
Suppose in the kth and p^{th} position of number n we have written 1. Then, all the remaining digits of n are zeros.

In this case, we have $n - 2 = \underbrace{9\ldots9}_{k-1} + \underbrace{9\ldots9}_{p-1}$. So $(n - 2) \vdots 3$.

The table shows all possible remainders when natural number m and m^2 are divided by 3.

(b)

m	0	1	2
m^2	0	1	1

From the table it is clear that $m^2 - 2$ is not divisible by 3, thus n cannot be a perfect square.

Problem 2.7: *Prove that $x^2 - y^2 = 2022$ has no natural number solution.*

Proof:
The table shows all possible remainders when natural number m and m^2 are divided by 4.

m	0	1	2	3
m^2	0	1	0	1

If x and y are natural numbers, then according to the table we have that either $x^2 - y^2$ is odd or it is divisible by 4. So, $x^2 - y^2 = 2022$ cannot have a natural solution, as 2022 is divisible by 2 but 2022 is not divisible by 4.

Problem 2.8: *Prove that $x^3 - y^3 = 2022$ has no natural number solution.*

Proof:
The table shows all possible remainders when natural number m and m^3 are divided by 9.

m	0	1	2	3	4	5	6	7	8
m^3	0	1	8	0	1	8	0	1	8

If x and y are natural numbers, then according to the table we have that $x^3 - y^3$ can leave only one of the following remainders 0, 1, 2, 7, 8 when divided by 9. Note that 2022 leave a remainder of 6 when divided by 9. So, $x^3 - y^3 = 2022$ has no natural number solution.

Problem 2.9: *What is the greatest number of consecutive natural numbers, so that their sum is equal to 2022?*

Answer: 12

Solution:
Let the sum of k consecutive positive integers be equal to 2022. Let n be the positive integer, so that the difference of the number of positive integers greater than n and the number of positive integers less than n is either equal to 0 or is equal to 1.
If k is odd, then
$$nk = 2022.$$
If k is even, then
$$nk + \tfrac{k}{2} = 2022.$$
Note that $2022 = 2 \cdot 3 \cdot 337$.

So, if k is odd, then its greatest possible value is 3.

If k is even, then from the following equation
$$k(2n+1) = 2^2 \cdot 3 \cdot 337,$$
we get that the greatest possible value of k is 12.

Note that if $k = 12$, then $n = 168$.

Problem 2.10: *A natural number is called "interesting" if it is possible to write a new number divisible by 7 by only using some of its digits, but not all of them. Find the greatest number that is not "interesting". (N.M. Sedrakyan)*

Answer: 999999

Solution:
Let us prove that any seven-digit number is "interesting". It is clear that any other natural number that has more than seven digits is "interesting" as well.

Proof by contradiction argument. Assume there exists a number $\overline{a_1a_2a_3a_4a_5a_6a_7}$ that is not "interesting" and none of its digits is 0. Consider numbers $a_7, \overline{a_6a_7}, \ldots, \overline{a_1a_2\ldots a_7}$. If we divide any two of these numbers by 7, then we get remainders different from each other (in another case the difference of any two of them would be divisible by 7). Thus, it follows that

$$\overline{a_1a_2\ldots a_7} \vdots 7.$$

Similarly, we get that

$$\overline{a_7a_1a_2\ldots a_6} \vdots 7.$$

From the last two results, we have that

$$\overline{a_1a_2\ldots a_7} - \overline{a_7a_1a_2\ldots a_6} \vdots 7.$$

On the other hand, we have

$$\overline{a_1\ldots a_7} - \overline{a_7a_1a_2\ldots a_6} = \overline{a_1\ldots a_6 0} + a_7 - \overline{a_7 000000} - \overline{a_1\ldots a_6} =$$

$$= 9 \cdot \overline{a_1\ldots a_6} - a_7 \cdot 999999,$$

and $999999 \vdots 7$. So $\overline{a_1\ldots a_6} \vdots 7$.
This is not possible and is a contradiction.
This ends the proof.
Note that 999999 is an "interesting" number.

Problem 2.11: *Find the number of natural numbers n such that $2^n + 5^n - 65$ is a perfect square. (N.M. Sedrakyan)*

Answer: 1
Solution:
Note that when $2, 2^2, 2^3, \ldots$ are divided by 5 the remainders are $2, 4, 3, 1, 2, 4, 3, 1, \ldots$.

It is clear that when n is odd, then when 2^n is divided by 5 the remainder is 2 or 3. So, when n is odd, then when $2^n + 5^n - 65$ is divided by 5 the remainder is 2 or 3.

Note that, when a perfect square is divided by 5 the remainder is 0, 1 or 4.

So, if n is odd, then $2^n + 5^n - 65$ cannot be a perfect square.

If n is even and is not less than 8, then we have

$$(5^{\frac{n}{2}})^2 < 2^n + 5^n - 65 < (5^{\frac{n}{2}} + 1)^2.$$

So $2^n + 5^n - 65$ is not a perfect square, as it is between two perfect squares.

Note that from $n = 2, 4, 6$ only for 4, we get that $2^n + 5^n - 65$ is a perfect square.

Problem 2.12: *Numbers* $1, 2, \ldots, 9$ *are written in a* 3×3 *board, such that in every square there is one number. Then the sum of each row, column, and diagonal was calculated. How many of the calculated sums can be prime?)*

Answer: 7
Solution:
Let us prove that at least one of the sums is divisible by 3 and is not less than 6, so the sum is not a prime number.
Let us consider an example, where seven of those sums are prime numbers, so the answer is 7.

2	1	8
9	3	5
6	7	4

Proof by contradiction argument.
Assume that none of those sums is divisible by 3. Suppose in the center square of the table we have the number a.

Let us subtract a from each number. Now, we have that the table is filled with consecutive natural numbers, in the center we have 0 and none of the new sums is divisible by 3.

Consider the sums that include 0. At least two of those sums do not include a number divisible by 3 (besides 0). So in each of those sums all addends (except 0) have the same remainder when divided by 3. Then, we have that addends of one or three of the sums are divisible by 3 or that when three addends are divided by 3 the remainders are (pairwise) different.

In that case, we have a sum divisible by 3. This is not possible and is a contradiction.
This ends the proof. So, the answer is 7.

Problem 2.13: *Find the product of natural numbers x, y, z, t such that $68(xyzt + xy + zt + xt + 1) = 157(yzt + y + t)$.*

Answer: 120
Solution:
We have
$$\frac{x(yzt+y+t)+zt+1}{yzt+y+t} = \frac{157}{68},$$
so
$$x + \frac{zt+1}{y(zt+1)+t} = 2 + \frac{21}{68}.$$

Note that
$$0 < \frac{zt+1}{y(zt+1)+t} < 1$$

thus $x = 2$ ($x \leq 1$ or $x \geq 3$ is not possible).

We have
$$\frac{y(zt+1)+t}{zt+1} = \frac{68}{21},$$

$$y + \frac{t}{zt+1} = 3 + \frac{5}{12}.$$

Then
$$y = 3 \text{ and } \frac{zt+1}{t} = \frac{21}{5}.$$

We get
$$z + \frac{1}{t} = 4 + \frac{1}{5}.$$

Thus $z = 4, t = 5$. So
$$xyzt = 2 \cdot 4 \cdot 4 \cdot 5 = 120.$$

Problem 2.14: *Find the product of prime numbers p, q and r such that $p(p-5) + q(q-5) = r(r+5)$.*

> Answer: 70
> **Solution**:
> The table shows the remainders when $m, m(m-5), m(m+5)$ are divided by 5.
>
> | m | 0 | 1 | 2 | 3 | 4 |
> |---|---|---|---|---|---|
> | m(m-5) | 0 | 1 | 4 | 4 | 1 |
> | m(m+5) | 0 | 1 | 4 | 4 | 1 |
>
> If none of the numbers p, q, r is 5, then after dividing the number $p(p-5) + q(q-5)$ by 5 we get one of the remainders $0, 2, 3$, which is not possible.
> If $p = 5$, then $q - 5 = r$. Thus $r = 2, q = 7$.
> If $q = 5$, then we have $p = 7, r = 2$.
> If $r = 5$, then $p(p-5) + q(q-5) = 50$. This is not possible.
> So $pqr = 70$

Problem 2.15: *Find the smallest possible value of n such that both n and $n + 2015$ are perfect squares. (N.M. Sedrakyan)*

> Answer: 289
> **Solution**:
> Let $n = a^2$ and $n + 2015 = b^2$, where a and b are natural numbers. We have $b^2 - a^2 = 2015$, so $(b-a)(b+a) = 2015$. We have $b - a$ is a divisor of 2015 and $b - a < b + a$. Thus $b - a < 45$.
> Note that $2015 = 5 \cdot 13 \cdot 31$, so $b - a \in \{1, 5, 13, 31\}$ and $b \in \{\frac{1}{2}(1 + 2015), \frac{1}{2}(5 + 403), \frac{1}{2}(13 + 155), \frac{1}{2}(31 + 65)\}$.
> The smallest possible value of b is 48 and the smallest possible value of n is $48^2 - 2015 = 289$.

Problem 2.16: *The sum of five natural numbers is 595. What is the greatest number of zeros that their product can end with? (N.M. Sedrakyan)*

Answer: 10
Solution:
Note that $595 = 5^3 + 5^3 + 2^5 \cdot 5 + 2^5 \cdot 5 + 5^2$, so the product $5^3 \cdot 5^3 \cdot 2^5 \cdot 5 \cdot 2^5 \cdot 5 \cdot 5^2$ ends with 10 zeros.
Let the sum of natural numbers a, b, c, d, e be 595. Let us prove that the product $abcde$ cannot end with 11 (or more) zeros.
Proof by contradiction argument. Assume that the product $abcde$ ends with 11 (or more) zeros (that means $(abcde) \vdots 10^{11}$) and let us show that it is not possible.

We can assume that $a \vdots 5, b \vdots 5, c \vdots 5, d \vdots 5, e \vdots 5$, as otherwise there are two of them which are not divisible by 5. So $(abc) \vdots 5^{11}$ and $d \not\vdots 5, e \not\vdots 5$. Thus $a \not\vdots 5^4$ or $b \not\vdots 5^4$ or $c \not\vdots 5^4$, which is not possible.

Let $a = 5x, b = 5y, c = 5z, d = 5t, e = 5u$, where $x, y, z, t, u \in \mathbb{N}$, then $x + y + z + t + u = 119$ and $(xyztu) \vdots 5^6 \cdot 2^{11}$. Note that two of the numbers x, y, z, t, u are divisible by 25, say $x \vdots 25$ and $y \vdots 25$. So $(x + y) \vdots 25$.

Consider the following three cases:

(a) $x + y = 50$, then $x = y = 25, z + t + u = 69, z \cdot t \cdot u \vdots 5^2 \cdot 2^{11}$ one of the numbers z, t, u is divisible by 2^6, which is not possible.

(b) $x + y = 75$, then $x = 25, y = 50$, (or $x = 50, y = 25$), $z + t + u = 44, (z \cdot t \cdot u) \vdots 5^2 \cdot 2^{10}$. Thus $z \vdots 16$. So $(t \cdot u) \vdots 25 \cdot 2^5$, which is not possible.

(c) $x + y = 100$, then $z + t + u = 19, z \cdot t \cdot u \vdots 5^2 \cdot 2^9$, which is not possible.

So the answer is 10.

Problem 2.17: *Find the smallest natural number that is divisible by 11 and the sum of its digits is 19.*

> Answer: 649
> **Solution**:
> The sum of the digits of the required number is 19. So, it can not be one-digit or two-digit number.
>
> Let us look for that number among three-digit numbers $\overline{abc}$.
> We know that $a + b + c = 19$ and
> $$\overline{abc} = 100a + 10b + c = (11(9a + b) + a - b + c) \vdots 11.$$
> So $(a - b + c) \vdots 11$, thus $(19 - 2b) \vdots 11$. Then $b = 4$ and $a + c = 15$, from which $a = 6$ and $c = 9$.

Problem 2.18: *Integers a, b, c are greater than 1 and have the following properties: $(a+b) \vdots (c+1)$, $(b+c) \vdots (a+1)$ and $(a+c) \vdots (b+1)$. Find the greatest possible value of $a + b + c$.*

> Answer: 59
> **Solution**:
> As the problem is symmetric with respect to a, b, c, then *without loss of generality* we can assume that $a \geq b \geq c$. In that case from the condition $b + c \leq 2a$ and $(b + c) \vdots (a + 1)$ we get $b + c = a + 1$.
> We have $a = b + c - 1$ and $(2c + b - 1) \vdots (b + 1)$, on the other hand $2c + b - 1 \leq 3b - 1$. So $2c + b - 1 = b + 1$ or $2c + b - 1 = 2(b + 1)$.
> When $2c + b - 1 = b + 1$, $c = 1$, $a = b$, which is not possible.
> When $2c + b - 1 = 2(b + 1)$, we get $b = 2c - 3$, $a = 3c - 4$ and $(5c - 7) \vdots (c + 1)$.
>
> On the other hand $12 = (5(c + 1) - (5c - 7)) \vdots (c + 1)$ and $a + b + c = 6c - 7$. Thus $a + b + c$ has the greatest possible value when c is the greatest.
>
> So $c = 11$ and the answer is 59.

Problem 2.19: *For numbers a, b we have: $(a,b)+[a,b] = a+b+1874$. Find the smallest possible value of $a + b$.*

> Answer: 941
> **Solution**:
> Let $(a,b) = d$, then $a = da_1, b = db_1$, where a_1 and b_1 are relatively prime. Thus, it follows that
> $$[a,b] = da_1b_1.$$
> We have
> $$d + da_1b_1 - da_1 - db_1 = 1874.$$
> This can be written as
> $$d(a_1 - 1)(b_1 - 1) = 2 \cdot 937.$$
> So $d = 1$ or $d = 937$.
>
> If $d = 1$, then $a_1 = 3, b_1 = 938$ or $a_1 = 938, b_1 = 3$. We get
> $$a + b = 941.$$
>
> If $d = 937$, then $a_1 = 3, b_1 = 2$ or $a_1 = 2, b_1 = 3$. We get
> $$a + b = 4685.$$
>
> So, the smallest possible value of $a + b$ is 941.

Problem 2.20: *Let m and n be natural numbers such that $(m+n, 10) = 1$. Prove that $\frac{m}{n}$ is possible to represent in the form*

$$\frac{10^{k-1}a_1 + 10^{k-2}a_2 + \ldots + a_k}{10^{k-1}b_1 + 10^{k-2}b_2 + \ldots + b_k},$$

where k is a natural number, $a_i, b_i \in \{1, 2, \ldots, 9\}$ and $a_i + b_i = 9$ for $i = 1, \ldots, k$.

Proof:
Let us first prove that if $(m+n, 10) = 1$, then there exists a unique nine-digit number that is divisible by $m+n$.

Let us consider numbers $9, 99, \ldots, \underbrace{9\ldots9}_{m+n+1}$.

Two of those $m+n+1$ numbers leave the same remainder when divided by $m+n$. Thus their positive difference is also divisible by $m+n$. Their positive difference is a number of the form $9\ldots90\ldots0$, where k and p are natural numbers. So $\underbrace{9\ldots9}_{m+n+1} \vdots (m+n)$.

Assume $9\ldots9 = ml + nl$, it is left to understand that there exist numbers $a_1, \ldots, a_k, b_1, \ldots b_k$ such that $ml = 10^{k-1}a_1 + \cdots + a_k$, $nl = 10^{k-1}b_1 + \cdots + b_k$ and $a_i + b_i = 9$, where $i = 1, \ldots, k$.

So we have that

$$\frac{m}{n} = \frac{ml}{nl} = \frac{10^{k-1}a_1 + \cdots + a_k}{10^{k-1}b_1 + \cdots + b_k}.$$

Problem 2.21: *A natural number is written in each square of 8×8 board. All numbers are pairwise different. It is known that any two numbers written in two neighboring squares are not relatively prime. Find the smallest possible value of the greatest of these numbers.*

Answer: 77
Solution:
Here is an example (a table) that satisfies the conditions of the problem.

19	57	27	3	9	12	39	13
38	76	24	72	18	8	26	65
56	70	54	44	6	32	64	52
28	49	63	33	48	16	4	58
7	21	77	11	22	36	2	62
14	35	55	66	40	75	60	74
46	42	10	15	20	50	68	34
23	69	45	5	25	30	51	17

If there exists a table with an answer smaller than 77, then note that numbers $1, 29, 31, 37, 41, 43, 47, 5, 59, 61, 67, 71, 73$ are not written in the table. So, the number of positive integers written in the table is not more than $76 - 13 = 63$, which is not possible.
So the answer is 77.

Problem 2.22: *Find the number of pairs (p, q), where p and q are primes, such that $p^4 - q$ and $q^4 - p$ are also prime.*

> Answer: 6
> **Solution**:
> Let $p \geq q$.
>
> We have $p^4 - q \geq q^4 - q = q(q^3 - 1) \geq 14$.
>
> If p and q are both odd, then $p^4 - q$ is even and from $p^4 - q \geq q^4 - q = q(q^3 - 1) \geq 14$ we have that $p^4 - q$ is not prime, which is not possible.
>
> Thus $q = 2$ and $p \in \{3, 5, 11, 13\}$.
>
> If $q = 2, p = 3$, then we get $2^4 - 3, 3^4 - 2$ (which are prime).
>
> If $q = 2, p = 5$, then we get $5^4 - 2 = 623$ (which is not prime, as $623 \vdots 7$).
>
> If $q = 2, p = 11$, then we get $2^4 - 11 = 5, 11^4 - 2 = 14639$ (which are prime).
>
> If $q = 2, p = 13$, then we get $2^4 - 13 = 3, 13^4 - 2 = 28559$ (which are prime).
>
> If $p \leq q$, then we get $p = 2, q = 3, p = 2, q = 11$ and $p = 2, q = 13$.

2.2 Problems from National Competitions

Problem 1: *A number was at first divided by 3, then by 6. The sum of two remainders is 7. Find the product of the remainders.*

Problem 2: *Let m, n be natural numbers such that $3n^2 - 2mn = 53$. Find the sum of m and n.*

Problem 3: *Find the smallest natural number that is not a divisor of $1 \cdot 2 \cdot 3 \cdots 88 \cdot 89$.*

Problem 4: *Find the smallest natural number n such that $\frac{n}{2}$ is a perfect square and $\frac{n}{3}$ is a perfect cube.*

Problem 5: *For numbers x, y, z we have $x + \frac{1}{y + \frac{1}{z}} = \frac{30}{13}$. Find z.*

Problem 6: *Find the smallest natural number different from 1, such that when divided by $4, 5, 6$ the remainders are 1.*

Problem 7: *When $\overline{ab} + \overline{ba}$ is divided by 9 the remainder is 5. What is the remainder when $\overline{ab}$ is divided by 9?*

Problem 8: *Let n be a natural number such that the least common multiple of $4, 9, n$ is 108. Find the smallest possible value of n.*

Problem 9: *Find the number of three digit numbers $\overline{abc}$, such that $\overline{cba}$ is a three digit number and $\overline{abc} + \overline{cba}$ is divisible by 5.*

Problem 10: *Find the remainder when 3^{100} is divided by 82.*

Problem 11: *Let x and y be the smallest natural number solutions of $3x^3 = 5y^2$. Find $x + y$.*

Problem 12: *James has more than 300 and less than 360 books. 40% of them are fiction books and $\frac{1}{11}$ of his books are poetry books. How many books does James have?*

Problem 13: *Find the sum of all natural values of n, such that $\frac{3n+1}{n+2}$ is an integer.*

Problem 14: *Given $x^2 + 3xy = 19$, where x and y are natural numbers. Find $x + y$.*

Problem 15: $\overline{24a8b}$ *is divisible by 5, 4 and 9. Find $a + b$.*

Problem 16: *How many zeros does the product $1 \cdot 2 \cdot \ldots \cdot 2020$ end with?*

Problem 17*: $\overline{abcd}$ *is divisible by $\overline{ab} \cdot \overline{cd}$. How many possible values of $\overline{abcd}$ are there?*

Problem 18: *Given that a is not an integer and $a + b, b + c, a + c$ are integers. Can $ab + bc + ac$ be an integer?*

Problem 19: *Are there natural numbers a and b, such that $ab(a+b)(a-b) = 2020$?*

Problem 20: *A natural number is written at each vertex of a square and on each side of this square is written the product of two numbers written at the endpoints of that side. The sum of all numbers written on the sides is equal to 2021. Find the sum of all numbers written at all vertices of this square.*

Problem 21: *Let a, b, c, d be natural numbers. Find the smallest value of $a + b + c$, such that for any d we have $(ad^b)^c = 64d^{12}$.*

Problem 22: *The sum of two natural numbers is 232. If we divide one of them by the other, the quotient is 3 and the remainder is 28. Find the product of these numbers.*

Problem 23: *Let the product of a three-digit number and all its digits be 2254. Find this three-digit number.*

Problem 24: *Find a three-digit number such that when divided by 37 its remainder is 31 and when divided by 38 its remainder is 12.*

Problem 25: *Two three-digit numbers were created using digits 1, 2, 3, 4, 5, 6. Find the smallest possible positive difference of those two numbers.*

Problem 26: *Seven years ago grandpa's age was a multiple of 8. Eight years ago it was a multiple of 9. How old is grandpa, if known that he was born in 20th century?*

2.2.1 Hints and Solutions

Problem 1: *A number was at first divided by 3, then by 6. The sum of two remainders is 7. Find the product of the remainders.*

> Answer: 10
> **Hint**: The remainders are 2 and 5.

Problem 2: *Let m, n be natural numbers such that $3n^2 - 2mn = 53$. Find the sum of m and n.*

> Answer: 132
> **Hint**: 53 is divisible by n.

Problem 3: *Find the smallest natural number that is not a divisor of $1 \cdot 2 \cdot 3 \cdots 88 \cdot 89$.*

> Answer: 97
> **Hint**: It is the smallest prime number greater than 89.

Problem 4: *Find the smallest natural number n such that $\frac{n}{2}$ is a perfect square and $\frac{n}{3}$ is a perfect cube.*

> Answer: 648
> **Hint**: $n = 2 \cdot 2 \cdot 2 \cdot 3 \cdot 3 \cdot 3 \cdot 3$.

Problem 5: *For numbers x, y, z we have $x + \frac{1}{y+\frac{1}{z}} = \frac{30}{13}$. Find z.*

> Answer: 4
> **Hint**: We have
> $$x + \frac{1}{y+\frac{1}{z}} = 2 + \frac{1}{\frac{13}{4}}.$$
> Thus
> $$x = 2, y + \frac{1}{z} = 3 + \frac{1}{4}.$$
> So $y = 3, z = 4$.

Problem 6: *Find the smallest natural number different from 1, such that when divided by $4, 5, 6$ the remainders are 1.*

> Answer: 61
> **Hint**: Let n be that number. We have $(n-1) \vdots 4, (n-1) \vdots 5, (n-1) \vdots 6$. Thus $(n-1) \vdots 60$.

Problem 7: *When $\overline{ab} + \overline{ba}$ is divided by 9 the remainder is 5. What is the remainder when $\overline{ab}$ is divided by 9?*

> Answer: 7
> **Hint**: $\overline{ab}$ and $\overline{ba}$ leave the same remainder when divided by 9.

Problem 8: *Let n be a natural number such that the least common multiple of $4, 9, n$ is 108. Find the smallest possible value of n.*

> Answer: 27
> **Hint**: $108 = 4 \cdot 27$.

Problem 9: *Find the number of three digit numbers $\overline{abc}$, such that $\overline{cba}$ is a three digit number and $\overline{abc} + \overline{cba}$ is divisible by 5.*

> Answer: 170
> **Hint**: $(a+c) \vdots 5$. Note that $4 \cdot 10 + 9 \cdot 10 + 4 \cdot 10 = 170$.

Problem 10: *Find the remainder when 3^{100} is divided by 82.*

> Answer: 81
> **Hint**: $3^{100} = (3^4)^{25} = (82-1)^{25} = 82 \cdot q - 1$, where q is a natural number.

Problem 11: *Let x and y be the smallest natural number solutions of $3x^3 = 5y^2$. Find $x + y$.*

> Answer: 60
> **Hint**: $x = 3 \cdot 5, y = 3 \cdot 3 \cdot 5$.

Problem 12: *James has more than 300 and less than 360 books. 40% of them are fiction books and $\frac{1}{11}$ of his books are poetry books. How many books does James have?*

> Answer: 330
> **Hint**: Let n be the number of books. We have $n \vdots 5$ and $n \vdots 11$. So $n \vdots 55$.

Problem 13: *Find the sum of all natural values of n, such that $\frac{3n+1}{n+2}$ is an integer.*

> Answer: -8
> **Hint**: Suppose $n+2 = m$. So $\frac{3n+1}{n+2} = 3 - \frac{5}{m}$. Note that $4 \cdot (-2) = -8$.

Problem 14: *Given $x^2 + 3xy = 19$, where x and y are natural numbers. Find $x + y$.*

> Answer: 7
> **Hint**: $x = 1, x + 3y = 19$.

Problem 15: *$\overline{24a8b}$ is divisible by $5, 4$ and 9. Find $a + b$.*

> Answer: 4
> **Hint**: $b = 0, a = 4$.

Problem 16: *How many zeros does the product $1 \cdot 2 \cdot \ldots \cdot 2020$ end with?*

> Answer: 503
> **Hint**: $1 \cdot 2 \cdot \ldots \cdot 2020 = 2^m \cdot 5^n \cdot a$, where m is equal to the sum of the number of numbers divisible by $2, 4, 2^{10} = 1024$ from $1, 2, \ldots, 2020$. Note that n is equal to the sum of the number of numbers divisible by $5, 25, 125, 625$ and $(a, 10) = 1$.

Problem 17: *$\overline{abcd}$ is divisible by $\overline{ab} \cdot \overline{cd}$. How many possible values of $\overline{abcd}$ are there?*

> Answer: 2
> **Hint**: Assume $\overline{abcd} = \overline{ab} \cdot \overline{cd} \cdot k$ where $k \in \mathbb{N}$. From here it follows $\overline{cd} = \frac{100}{k \cdot \overline{ab}} \cdot \overline{ab}$. So, from $(\overline{ab}, k \cdot \overline{ab} - 1) = 1$ we get $100 \vdots (k \cdot \overline{ab} - 1)$.

Problem 18: *Given that a is not an integer and $a + b, b + c, a + c$ are integers. Can $ab + bc + ac$ be an integer?*

> Answer: It is not possible
> **Hint**: $ab + bc + ac = (a + b)(a + c) - a^2$.

Problem 19: *Are there natural numbers a and b, such that $ab(a+b)(a-b) = 2020$?*

> Answer: There are no such natural numbers.
> **Hint**: $2020 = 2 \cdot 2 \cdot 5 \cdot 101$, thus $a + b \geq 101$ and $ab(a+b)(a-b) \geq 51 \cdot 101$.

Problem 20: *A natural number is written at each vertex of a square and on each side of this square is written the product of two numbers written at the endpoints of that side. The sum of all numbers written on the sides is equal to 2021. Find the sum of all numbers written at all vertices of this square.*

> Answer: 90
> **Hint**: Let numbers a, b, c, d be written on the vertices of a square. According to the condition of the problem we have $ab + bc + cd + da = 2021$. Thus, $(a+c)(b+d) = 2021 = 43 \cdot 47$. So $(a+c) + (b+d) = 43 + 47 = 90$.

Problem 21: *Let a, b, c, d be natural numbers. Find the smallest value of $a + b + c$, such that for any d we have $(ad^b)^c = 64d^{12}$.*

> Answer: 10
> **Hint**: $d^{bc-12} = \frac{64}{a^c}$. So $bc - 12 = 0$ and $a^c = 64$. We have $b + c \geq 7$. When $a = 2, c = 6, b = 2$, then $a + b + c = 10$. When $a \geq 4$, then $a + b + c \geq 11$.

Problem 22: *The sum of two natural numbers is 232. If we divide one of them by the other, the quotient is 3 and the remainder is 28. Find the product of these numbers.*

> Answer: 9231

Problem 23: *Let the product of a three-digit number and all its digits be 2254. Find this three-digit number.*

> Answer: 322
> **Hint**: $\overline{abc} \cdot (a + b + c) = 2 \cdot 7 \cdot 7 \cdot 23$. Thus $\overline{abc} \vdots 23$. Note that $322 = 23 \cdot 14$.

Problem 24: *Find a three-digit number such that when divided by 37 its remainder is 31 and when divided by 38 its remainder is 12.*

> Answer: 734
> **Hint**: $n = 38k + 12$, where $k \in \{3, 4, \ldots, 25\}$. Thus, $n = 37k + (k + 12)$, therefore $k + 12 = 31$.

Problem 25: *Two three-digit numbers were created using digits 1, 2, 3, 4, 5, 6. Find the smallest possible positive difference of those two numbers.*

> Answer: 47
> **Hint**: $\overline{abc} - \overline{def} = 100(a - d) - (\overline{ef} - \overline{bc}) \geq 100 - (\overline{ef} - \overline{bc})$. We have $\overline{ef} \leq 65, \overline{bc} \geq 12$. Thus, $\overline{ef} - \overline{bc} \leq 53$, and $\overline{abc} - \overline{def} \geq 47$.

Problem 26: *Seven years ago grandpa's age was a multiple of 8. Eight years ago it was a multiple of 9. How old is grandpa, if known that he was born in 20th century?*

> Answer: 71
> **Hint**: Let n be the age of the grandpa. As $(n-7) \vdots 8$ and $(n-8) \vdots 9$, then $(n+1) \vdots 8$ and $(n+1) \vdots 9$, thus $(n+1) \vdots 72$.
> So $n + 1 = 72$ (as he was born in 20th century). Thus $n = 71$.

Chapter 3

Pigeonhole principle

We define **pigeonhole principle** by the following statement:

Consider n *boxes* (holes) and $n \cdot k + 1$ *pigeons*, where n and k are natural numbers. If we want to place all pigeons into these boxes, then there is a box that contains at least $k + 1$ pigeons.

Another (geometric) statement of the pigeonhole principle is this one:

If the sum of the lengths of some segments on a line of length l is greater than $k \cdot l$, where k is a natural number, then at least $k + 1$ of these segments intersect.

Let us consider some examples where these statements can be used.

Example 3.1:

There are 13 students in 7th grade. Prove that there are two students who were born in the same month.

Example 3.1: Proof
Proof: There are 12 months in a year and the number of students is 13. We have that $13 = 12 \cdot 1 + 1$. According to the pigeonhole principle two of the students celebrate their birthdays in the same month.

> **Example 3.2:**
>
> *Prove that from any four integers it is possible to choose two, such that their difference is divisible by 3.*

> **Example 3.2: Proof**
> **Proof:** The possible remainders after dividing an integer by 3 are $0, 1$ and 2. Let us call these numbers *boxes* (holes) and four integers *pigeons*. We have four pigeons in three boxes. According to the pigeonhole principle, one of the boxes includes at least two pigeons. From here it follows that dividing two of those integers by 3 result in the same remainder. Suppose these numbers are a and b. We have $a = 3 \cdot q_1 + r$ and $b = 3 \cdot q_2 + r$, where q_1, q_2 are integers, and $r \in \{0, 1, 2\}$. So $a - b = 3q_1 + r - (3q_2 + r) = 3(q_1 - q_2)$. Thus $(a - b) \vdots 3$

3.1 Practice Problems

Problem 3.1. *Prove that in any group of six people, there are two people who know the same number of people (from that group).*

Problem 3.2. *11 numbers were chosen from 1, 2,..., 20. Prove that there are at least two numbers among these 11 numbers, such that one of them is divisible by the other one.*

Problem 3.3. *675 numbers were chosen from 1,2,..., 2022. Prove that there are numbers a and b among these 675 numbers, where $a > b$, such that $a + b$ is divisible by $a - b$.*

Problem 3.4. *Prove that it is possible to multiply 2023 by a natural number so that all the digits of the product are ones.*

Problem 3.5. *Consider any 51 points inside of a square of side length 5. Prove that at least three of these points can be covered by a unit square (of side length 1).*

Problem 3.6. *Prove that from any five integers (written on the same line) one of them is divisible by 5 or it is possible to choose some consecutive numbers such that their sum is divisible by 5.*

Problem 3.7. *Hotel rooms are numbered from 1 to 11. It is known that in each room live one or two people. Prove that there exist rooms with consecutive numbers where live exactly eleven people.*

Problem 3.8. *How many knights is possible to place on a 4×4 chessboard such that any knight hits no more than two others?*

Problem 3.9. *The figure in the picture is made of four identical (same) squares and can be covered by five identical squares. Prove that it can be covered by four of these five identical squares.*

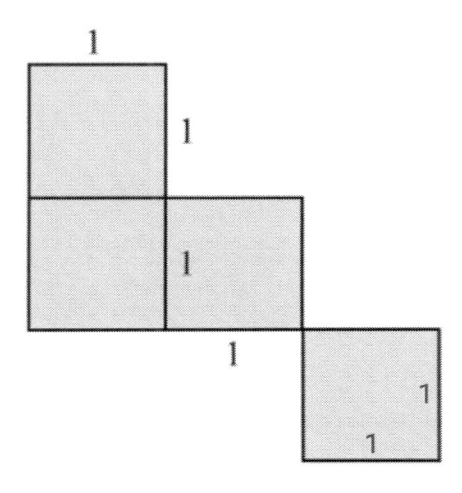

Problem 3.10. *Inside a square of side length 2 there are four squares such that the sum of their perimeters is 16. Prove that there exists a line that crosses (intersects) at least three of these four squares.*

Problem 3.11. *Any 33 squares of a chessboard were chosen. On each of these (chosen) squares there is a rook. Prove that there are five rooks that do not hit each other.*

3.1.1 Solutions of practice problems

Problem 3.1: *Prove that in any group of six people, there are two people who know the same number of people (from that group).*

> **Proof**: Note that for each person (in a group of six people) the number of friends (in that group) cannot be more than five.
>
> If two people know each other, then let us call them "friends". Let us consider the number of friends (in that group) for each of these six people. So, there are six different options for each of them (they can have 0, 1, 2, 3, 4, 5 friends in that group). Note that 0 and 5 cannot be among these six numbers at the same time, as if one of the people does not have any friends (in that group) then there is no person (in that group) who is a friend with that person (so the number of friends of this person cannot be 5), which means 0 and 5 cannot be among these six numbers at the same time.
>
> Thus, we have six integers which are the *pigeons*, where the number of integers that are different from each other is not more than five (these are the *boxes* (holes)). According to the pigeonhole principle, at least two people from these six people have the same number of friends in that group.

Problem 3.2: *11 numbers were chosen from 1, 2,..., 20. Prove that there are at least two numbers among these 11 numbers, such that one of them is divisible by the other one.*

> **Proof**: Consider the following "number chains": $1-2-4-8-16, 3-6-12, 5-10-20, 7-14, 9-18, 11, 13, 15, 17, 19$. Call these "number chains" *boxes* (holes) and call 11 chosen numbers *pigeons*. We have that 11 *pigeons* are in 10 *boxes* (holes). According to the pigeonhole principle there are at least two *pigeons* in one of the *boxes* (holes). Note that if the numbers are in the same "number chain", then one of them is divisible by the other one.

Problem 3.3: *675 numbers were chosen from 1,2,..., 2022. Prove that there are numbers a and b among these 675 numbers, where $a > b$, such that $a + b$ is divisible by $a - b$.*

> **Proof**: Let the *boxes* (holes) be the following triples: (1,2,3), (4,5,6), ..., (2020,2021,2022) and the *pigeons* be the chosen 675 numbers.
>
> The number of boxes is 2022 : 3 = 674. According to the Pigeonhole principle, there are at least two pigeons in one of the boxes. That is, there exists $n \in \{1, 4, 7, \ldots, 2020\}$ such that at least two of the numbers $n, n+1, n+2$ are chosen.
>
> Note that the greatest number is a and the smallest number is b, as $((n+1)+n) \vdots ((n+1)-n), ((n+2)+n) \vdots ((n+2)-n)$ and $((n+2)+(n+1)) \vdots ((n+2)-(n+1))$.

Problem 3.4: *Prove that it is possible to multiply 2023 by a natural number so that all the digits of the product are ones.*

> **Proof**: Let the numbers $1, 11, 111, \ldots, \underbrace{11\ldots1}_{2024 \text{ times}}$ be the *pigeons* and $0, 1, \ldots, 2022$ be the *boxes* (holes) or the numbers of the *boxes*.
>
> Possible remainders after dividing a number by 2023 are $0, 1, \ldots, 2022$. We say that a number is in box i if after dividing that number by 2023 the remainder is i. So, we have that 2024 *pigeons* are in 2023 *boxes*. According to the pigeonhole principle, two of them are in the same *box*.
>
> We got that there are $\underbrace{1\ldots1}_{m \text{ times}}$ and $\underbrace{1\ldots1}_{n \text{ times}}$ ($m > n$) numbers, that give the same remainder when divided by 2023.
> So their difference $\underbrace{1\ldots1}_{m-n \text{ times}} \underbrace{0\ldots0}_{n \text{ times}}$ is divisible by 2023.
> So $\underbrace{1\ldots1}_{m-n \text{ times}}$ is divisible by 2023.

Problem 3.5: *Consider any 51 points inside of a square of side length 5. Prove that at least three of these points can be covered by a unit square (of side length 1).*

Proof: Let us divide this square into 25 unit squares (see the picture).

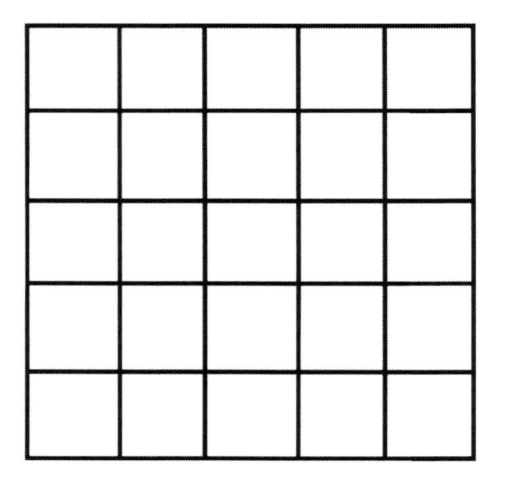

Let these squares be the *boxes* and let 51 points be the *pigeons*. If the point belongs to more than one unit square, then we consider that it belongs only to one of them (does not matter which one). We have 51 pigeons are in 25 boxes and $51 = 25 \cdot 2 + 1$. According to the pigeonhole principle, at least three of those *pigeons* belong to one square.

So, at least three of these 51 points is possible to cover by a unit square.

Problem 3.6: *Prove that from any five integers (written on the same line) one of them is divisible by 5 or it is possible to choose some consecutive numbers such that their sum is divisible by 5.*

> **Proof**: Let these five integers be a, b, c, d, e.
> Consider the following five integers:
> $a, a+b, a+b+c, a+b+c+d$ and $a+b+c+d+e$.
> If one of these numbers is divisible by 5, then this ends the proof.
>
> If none of the numbers $a, a+b, a+b+c, a+b+c+d, a+b+c+d+e$ is divisible by 5, then when divided by 5 they leave one of the following remainders $1, 2, 3, 4$.
>
> According to the Pigeonhole principle, when divided by 5 two of these numbers leave the same remainder: for example $a+b$ and $a+b+c+d$. In that case, their difference is divisible by 5. Note that $a+b+c+d-(a+b) = c+d$ and this means that $c+d$ is divisible by 5. So, this ends the proof.

Problem 3.7: *Hotel rooms are numbered from 1 to 11. It is known that in each room live one or two people. Prove that there exist rooms with consecutive numbers where live exactly eleven people.*

> **Proof**: According to the solution of problem 6, there exist rooms with consecutive numbers such that the number of people living there is a multiple of 11. From the condition of the problem, it is clear that this number is 11 or 22. It cannot be 22, because we get there are two people living in each room, which is not possible. So, in those rooms with consecutive numbers live exactly eleven people.

Problem 3.8: *How many knights is possible to place on a 4×4 chessboard such that any knight hits no more than two others?*

Answer: 12
Solution: Here is an example that satisfies the conditions of the problem.

	•	•	•	•
	•			•
	•			•
	•	•	•	•

Now let us prove that if there are 13 knights on a 4×4 chessboard, then one of them will definitely hit at least three other knights. Enumerate the squares of a 4×4 chessboard with numbers $1, 2, 3, 4$ (see the picture below).

3	4	3	1
2	1	2	4
4	3	4	2
1	2	1	3

Consider numbers $1, 2, 3, 4$ as the *boxes* (holes) and there are not less than 13 knights (*pigeons*) in these *boxes*. According to the pigeonhole principle, there is a *box* that contains at least four knights. Thus, there are four boxes with the same number and in each of them there is a knight. It is left to see that one of these knights hits the other three, which is not possible.

Problem 3.9: *The figure in the picture is made of four identical (same) squares and can be covered by five identical squares. Prove that it can be covered by four of these five identical squares.*

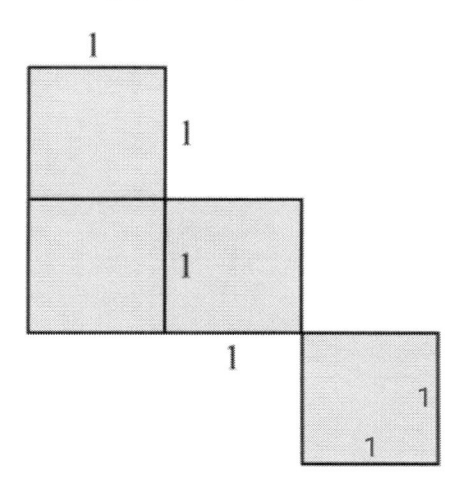

Proof: Consider the following six points (see the picture).

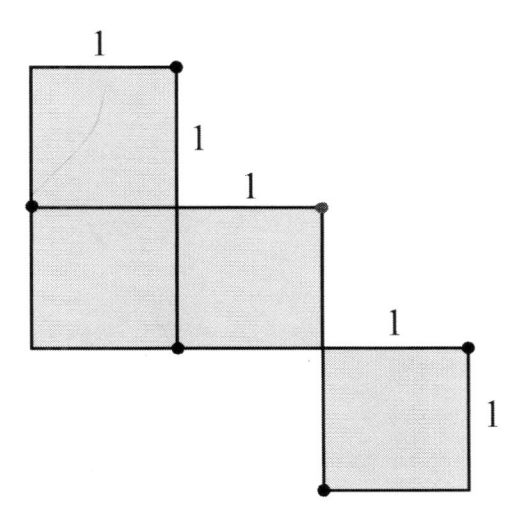

Note that for any two of these points their distance is not less then the diagonal of the unit square.

According to the Pigeonhole principle, there is square among these five squares that at least two of these six points belong to.

Thus, the sides of those five squares are not less than 1. So, it is possible to cover the figure by four of these five squares.

Problem 3.10: *Inside a square of side length 2 there are four squares such that the sum of their perimeters is 16. Prove that there exists a line that crosses (intersects) at least three of these four squares.*

Proof: Consider the "perpendicular projections" of those squares on side AB of the square of side length 2 (see the picture below).

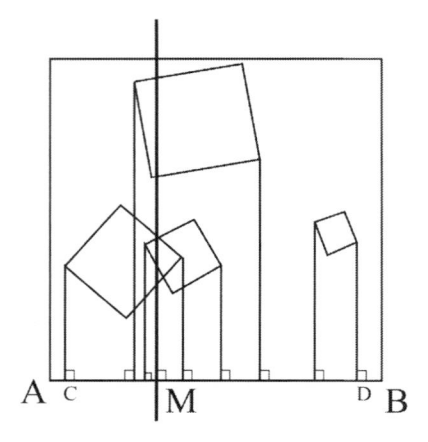

Let us consider the figure below. We have $d \geq m$ and $m \geq a$, thus $d \geq a = \frac{1}{4} \cdot 4a$.

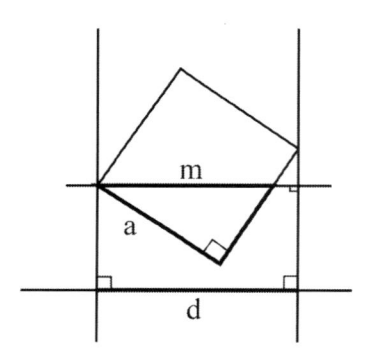

So, on CD we have four line segments with lengths less than 2, where the sum of their lengths is not less than $\frac{1}{4} \cdot 16 = 4$.

According to the Pigeonhole principle, there is a point M that belongs to at least three of these projections. In that case the line passing through M and perpendicular to AB crosses (intersects) at least three of the squares which are in the square of side length 2.

Problem 3.11: *Any 33 squares of a chessboard were chosen. On each of these (chosen) squares there is a rook. Prove that there are five rooks that do not hit each other.*

> **Proof**: Consider the columns of the chessboard as squares. There are 33 rooks in 8 squares. According to the Pigeonhole principle, in one of the squares there are at least five rooks, as $33 = 8 \cdot 4 + 1$.
>
> Do not consider the column that contains at least five rooks. In the remaining columns there are at least $33 - 8 = 25$ rooks. According to the pigeonhole principle, in one of those seven columns there are at least four rooks. Similarly, there is a column that contains at least three rooks. In this manner there is the fourth column with at least two rooks and the fifth column with at least one rook.
>
> Choose a rook in the fifth column. We can now choose a rook in the fourth column that is not on the same line with the rook chosen from the fifth column. Similarly choose one in the third column, one in the fourth column and one in the fifth column.
>
> These five chosen rooks are in different columns and rows, thus there are no rooks that hit each other. This ends the proof.

3.2 Problems from National Competitions

Problem 1: *There is a box in a dark room. The box contains 10 white, 13 red, 15 blue and 12 green balls. What is the smallest number of balls to take out from the box in order one of them to be green?*

Problem 2: *There are 101 rectangles of natural number side lengths less than 101. Prove that it is possible to choose three rectangles such that the first rectangle can be covered by the second one, and the second by the third one. (N. M. Sedrakyan)*

Problem 3: *13 points divide a circle into 13 equal parts. Each of these 13 points is labeled by one of the numbers $1, 2, 3$. Prove that it is possible to choose three points from these 13 points such that they are labeled by the same number and they form an isosceles triangle.*

Problem 4: *From a square of side length 4 three unit squares (side length of 1) were cut. Prove that from the rest of the paper is possible to cut one more unit square.*

3.2.1 Hints and Solutions

Problem 1: *There is a box in a dark room. The box contains 10 white, 13 red, 15 blue and 12 green balls. What is the smallest number of balls to take out from the box in order one of them to be green?*

> Answer: 39
> **Hint**: Consider white, red and blue balls as one color (for example yellow).

Problem 2: *There are 101 rectangles of natural number side lengths less than 101. Prove that it is possible to choose three rectangles such that the first rectangle can be covered by the second one, and the second by the third one. (N. M. Sedrakyan)*

> **Hint**: Consider the point (a, b) on a coordinate system instead of rectangle $a \times b$, where $a \leq b$.

Problem 3: *13 points divide a circle into 13 equal parts. Each of these 13 points is labeled by one of the numbers $1, 2, 3$. Prove that it is possible to choose three points from these 13 points such that they are labeled by the same number and they form an isosceles triangle.*

> **Hint**: According to the pigeonhole principle five of these 13 points are labeled by the same number.

Problem 4: *From a square of side length 4 three unit squares (side length of 1) were cut. Prove that from the rest of the paper is possible to cut one more unit square.*

Hint: Let points A and B be any points in the unit square (see the drawing).

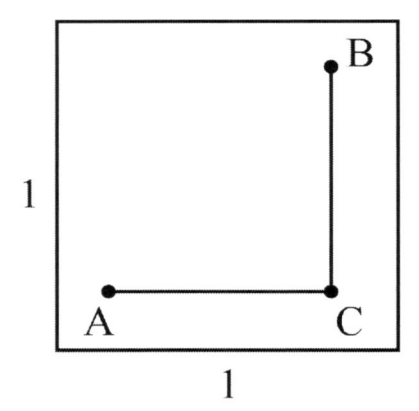

By triangle inequality we have $AB \leq AC + BC \leq 2$.
So the squares which were cut may have a common part with (and "destroy") at most one black square, as $AB \leq 2$ (see the drawing).

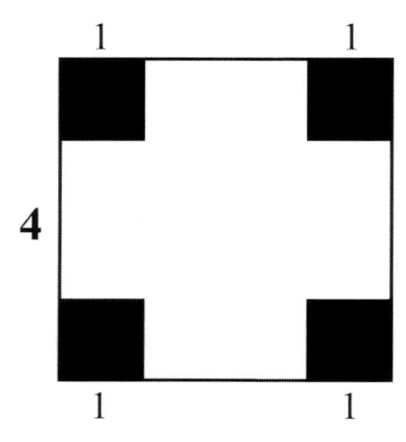

Chapter 4

Combinatorics

The main goal of **combinatorics** is to help to count the number of **elements** (members) of some **set** (group).

Mostly, the simplest way to do this is to write down all *elements* and count them. It is important not to write the same *element* more than once and not to miss any *element*.

The *rule of addition* (also called the *addition principle*) and the *rule of multiplication* (also called the *multiplication principle*) are fundamental principles of combinatorics.

Addition Principle. Suppose it is possible to choose the first object in m ways and the second object in n ways. In this case, exactly one object from the first or second objects is possible to choose in $m + n$ ways.

Multiplication Principle. Suppose it is possible to choose the first object in m ways, after which it is possible to choose the second object in n ways. In this case, the ordered pair of first and second objects is possible to choose in $m \cdot n$ ways.

Addition and *multiplication principles* can be generalized to used for more than two objects.

Example 4.1:
Find the number of all two-digit numbers.

Example 4.1: Solution
Answer: 90
Solution: There are 9 ways to choose the first digit of a two-digit number. The number of ways to choose the second digit is 10. According to the *multiplication principle* the number of two-digit numbers is $9 \cdot 10 = 90$.

Example 4.2:
How many three-digit numbers can be formed by three different digits?

Example 4.2: Solution
Answer: 648
Solution: There are 9 ways to choose the *hundreds digit*, $10 - 1 = 9$ ways to choose the *tens digit*, and $10 - 2 = 8$ ways to choose the *units digit*. According to the *multiplication principle* the number of three-digit numbers with different digits is $9 \cdot 9 \cdot 8 = 648$.

Example 4.3:
Consider a three-digit number which is formed by three different digits. The sum of its first two digits is equal to 9 and its last digit is not 0. How many such three-digit numbers are there?

Example 4.3: Solution
Answer: 64
Solution: According to the condition of the problem, the first two digits can be 1 and 8, 2 and 7, 3 and 6, 4 and 5, and 0 and 9. For the first four cases there are 7 ways to choose the third digit and for the last case there are 8 ways.
According to *multiplication* and *addition principles* the number of such three-digit numbers is $8 \cdot 7 + 8 = 64$.

Example 4.4:

The nine vertices of a 2×2 square are colored in blue or red colors. The bottom left and upper right vertices have different colors. Also, for any two vertices of the same color there is a road, consisting of the sides of the squares, that passes through the vertices of the same color connecting these two vertices. How many possible ways are there to color these nine vertices?

Example 4.4: Solution

Answer: 60

Solution: Suppose that the bottom left vertex is blue and the upper right one is red (see the picture).

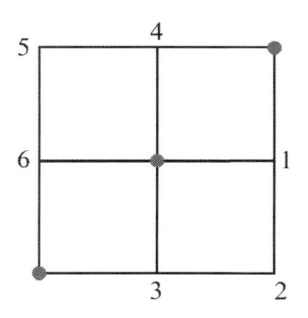

Consider the case when the middle vertex of the square is blue. If one of the vertices $1, 2, 3$ is red then any vertex that has a number less than that vertex is also red. The same statement is true for the vertices $4, 5, 6$. So the number of colorings that satisfy the assumption of the problem is $1 + 2 + 3 + 4 + 3 + 2 = 15$.

So the overall number of colorings is $2 \cdot 2 \cdot 15 = 60$.

4.1 Practice Problems

Problem 4.1. *How many ways are there to arrange five different books (labeled by the numbers 1, 2, 3, 4, 5) on a bookshelf so that book number 2 is between book number 1 and book number 3?*

Problem 4.2. *How many possible ways are there to travel from city A to city B and back to city A, if from A to B is possible to travel only in one direction and from B to A only in the opposite direction? The road consists of several parts labeled by **different** numbers, for example one way to go from A to B and come back to A is $1, 4, 8, 10, 5, 2$ (see the picture). The road should not include the same number more than once, for example $1, 4, 8, 8, 5, 2$ does not work.*

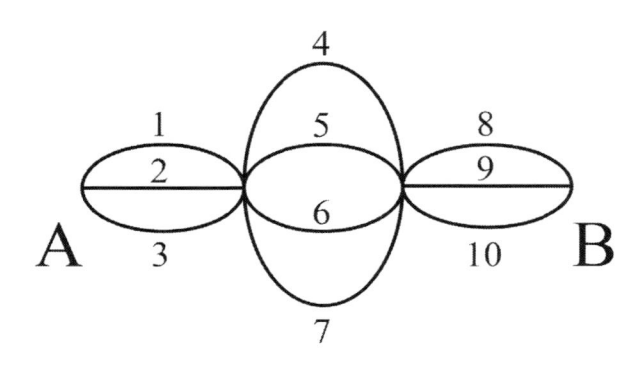

Problem 4.3. *Find the number of all three-digit numbers so that for each of them the sum of its digits is divisible by 5.*

Problem 4.4. *Find the number of five-digit numbers so that each of them is written using the digits $1, 2, 3, 4, 5$ (each digit is used once). Moreover, 1 is on the left side of 2 and 2 is on the left of 3, for example $51243, 15243$.*

Problem 4.5. *Find the number of three-digit numbers $\overline{abc}$, so that a, b, c are different from each other and $a > b$, $c > b$.*

Problem 4.6. *Find the number of positive divisors of 2520.*

Problem 4.7. *Find the number of positive odd divisors of 88200.*

Problem 4.8. *In how many ways is it possible to cover 2×5 rectangle by five 1×2 rectangles? The order does not matter.*

Problem 4.9. *The figure below is made of 14 unit squares. In how many ways is it possible to cover this figure by seven 1×2 rectangles?*

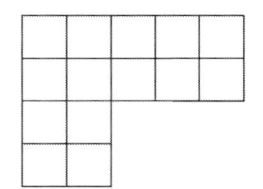

Problem 4.10. *A four-digit number is called "interesting", if it consists of two two-digit numbers of different parities, for example 1724, 5211. How many four-digit "interesting" numbers are there?*

Problem 4.11. *A four-digit number is called "ordinary", if it has at least two neighbor digits of different parities, for example 5200, 5211. How many four-digit "ordinary" numbers are there?*

Problem 4.12. *A five-digit number divisible by 11 is called "beautiful", if its digits are different from each other and are 1, 2, 3, 4, 8 (in some order). Find the number of all five-digit "beautiful" numbers.*

Problem 4.13. *What is the greatest number of three-digit numbers so that all of them have the same sum of the digits? For example there are **three** three-digit numbers so that the sum of the digits of each of them is equal to 2 (that is 101, 110, 200) and there are **five** three-digit numbers so that the sum of the digits of each of them is equal to 3 (that is 102, 111, 120, 201, 210). **Five** is greater than **three**, but five is not the greatest possible answer. So, what is the greatest possible answer?*

Problem 4.14. *Suppose for numbers a, b, c, d we have $\overline{ab}, \overline{cd}, \overline{ac}, \overline{bd}$ and $\overline{ab} + \overline{cd} = \overline{ac} + \overline{bd}$. Find the number of all possible (a, b, c, d).*

Problem 4.15. *In how many ways is it possible to put five candies in three pots, if we know that in the first pot is possible to put not more than one candy, in the second pot not more than two candies and in the third pot not more than three candies?*

Problem 4.16. *In how many ways is it possible to put different books on a bookshelf if the first, second and third books should not be next to the fourth book?*

Problem 4.17. *Four points were chosen on each edge of a triangular pyramid. Consider those 24 points and four vertices of the pyramid. Find the number of lines that pass through two of the considered 28 points.*

Problem 4.18. *The squares of a 3×3 board are colored in red, blue or yellow. Any two squares that share a side are colored in different colors. In how many such ways is it possible to color this board? (N.M. Sedrakyan)*

Problem 4.19. *Find the number of positive divisors of 770^7 so that when divided by 3 the remainder of each of them is 1.*

Problem 4.20. *There are six identical (same) mathematics books and four different physics books. In how many ways is it possible to put these books on a bookshelf so that one neighbor of a physics book is a mathematics book and the other one is a physics book?*

4.1.1 Solutions of practice problems

Problem 4.1: *How many ways are there to arrange five different books (labeled by the numbers 1, 2, 3, 4, 5) on a bookshelf so that book number 2 is between book number 1 and book number 3?*

> Answer: 240
> **Solution:** Consider the first three volumes as one book. Now we need to arrange five books on a shelf. The number of possibilities for all arrangements is 5!. Note that the possible arrangements of the first three volumes are I, II, III or III, II, I. Thus the overall number of arrangements is $2 \cdot 5! = 240$.

Problem 4.2: *How many possible ways are there to travel from city A to city B and back to city A, if from A to B is possible to travel only in one direction and from B to A only in the opposite direction? The road consists of several parts labeled by **different** numbers, for example one way to go from A to B and come back to A is 1, 4, 8, 10, 5, 2 (see the picture). The road should not include the same number more than once, for example 1, 4, 8, 8, 5, 2 does not work.*

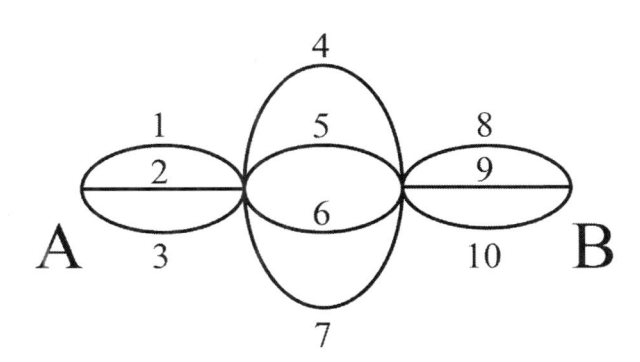

> Answer: 432
> **Solution:** The number of possible roads from A to B is $3 \cdot 4 \cdot 3$. The number of roads that come back is $2 \cdot 3 \cdot 2$. Thus the number of ways to travel from A to B and back is $36 \cdot 12 = 432$.

Problem 4.3: *Find the number of all three-digit numbers so that for each of them the sum of its digits is divisible by 5.*

> Answer: 180
> **Solution**: Suppose the sum of the digits of $\overline{abc}$ is divisible by 5. Note that it is possible to choose a in 9 ways. After choosing a it is possible to choose b in 10 ways. Suppose when $a + b$ is divided by 5 the remainder is r. Note that exactly two of the numbers $0, 1, \ldots, 9$ will leave a remainder $5 - r$ when divided by 5 (when $r = 0$ we say 0 instead of $5 - r$). According to the *multiplication principle*, the number of such three-digit numbers is $9 \cdot 10 \cdot 2 = 180$.

Problem 4.4: *Find the number of five-digit numbers so that each of them is written using the digits $1, 2, 3, 4, 5$ (each digit is used once). Moreover, 1 is on the left side of 2 and 2 is on the left of 3, for example $51243, 15243$.*

> Answer: 20
> **Solution**: Let us write the numbers $1, 2, 3$ from left to right. There are 4 possibilities of writing those numbers $(4123, 1423, 1243, 1234)$. After writing these 4 numbers we can add 5 possibilities of writing number 5. According to the *multiplication principle*, the number of five-digit numbers satisfying the problem is $4 \cdot 5 = 20$.

Problem 4.5: *Find the number of three-digit numbers $\overline{abc}$, so that a, b, c are different from each other and $a > b$, $c > b$.*

> Answer: 240
> **Solution**: When $b = 0$, then there are nine possibilities of choosing 9. After choosing a there are eight possibilities of choosing c. For $b = 0$ there are $9 \cdot 8 = 72$ numbers satisfying the conditions of the problem.
> Similarly, for $b = 1, b = 2, \ldots, b = 7$ the number of satisfying numbers are $8 \cdot 7, 7 \cdot 6, \ldots, 2 \cdot 1$.
> According to the *addition principle*, the number of overall numbers satisfying the conditions of the problem is $9 \cdot 8 + 8 \cdot 7 + \cdots + 2 \cdot 1 = 240$.

Problem 4.6: *Find the number of positive divisors of 2520.*

Answer: 48

Solution: Let us represent 2520 as a product of prime numbers.
We have $2520 = 2 \cdot 2 \cdot 2 \cdot 3 \cdot 3 \cdot 5 \cdot 7 = 2^3 \cdot 3^2 \cdot 5 \cdot 7$
Note that every positive divisor of 2520 has the form $2^a \cdot 3^b \cdot 5^c \cdot 7^d$, where $a \in \{0, 1, 2, 3\}, b \in \{0, 1, 2\}, c \in \{0, 1\}, d \in \{0, 1\}$.
According to the fundamental theorem of arithmetic, if (a, b, c, d) are different from each other, then $2^a \cdot 3^b \cdot 5^c \cdot 7^d$ are different. Thus, it follows that the number of positive divisors of 2520 is the number of 4-tuples (a, b, c, d), where $a \in \{0, 1, 2, 3\}, b \in \{0, 1, 2\}, c \in \{0, 1\}, d \in \{0, 1\}$.
According to the *multiplication principle*, the number of such 4-tuples is

$$4 \cdot 3 \cdot 2 \cdot 2 = 48.$$

Problem 4.7: *Find the number of positive odd divisors of 88200.*

Answer: 27

Solution: Represent 88200 in canonical form.

$$88200 = 2^3 \cdot 3^2 \cdot 5^2 \cdot 7^2.$$

Each odd divisor of 88200 has the form $3^a \cdot 5^b \cdot 7^c$, where $a, b, c \in \{0, 1, 2\}$. Thus, we need to find the number of triplets (a, b, c), where $a, b, c \in \{0, 1, 2\}$ (see the solution of problem 4.6). According to the *multiplication principle*, that number equals $3 \cdot 3 \cdot 3 = 27$.

Problem 4.8: *In how many ways is it possible to cover 2×5 rectangle by five 1×2 rectangles? The order does not matter.*

> Answer: 8.
>
> **Solution**: Note that after covering the bottom left square of a 2×5 lattice with a 1×2 rectangle we will either need to cover 2×4 rectangle with 4 rectangles with 2 squares, or we need to cover 2×3 rectangle with 3 rectangles that have 2 squares (see the picture). 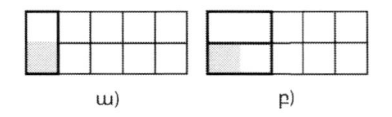
>
> m) p)
>
> From the facts above we can conclude that rectangles of size $2 \times 1, 2 \times 2, 2 \times 3, 2 \times 4, 2 \times 5$ can be covered with rectangles consisting of 2 squares in $1, 2, 1 + 2 = 3, 2 + 3 = 5, 3 + 5 = 8$ ways.

Problem 4.9: *The figure below is made of 14 unit squares. In how many ways is it possible to cover this figure by seven 1×2 rectangles?*

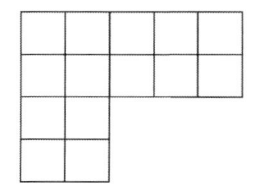

> Answer: 19
>
> **Solution**: Note that after covering the bottom left 2×2 square we need to cover one of the following figures (see the picture). According to the solution of problem 4.8 the answer is $8 + 3 + 8 = 19$.
>
> m) p) q)

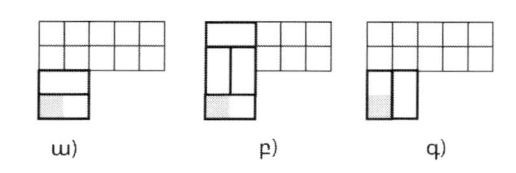

Problem 4.10: *A four-digit number is called "interesting", if it consists of two two-digit numbers of different parities, for example 1724, 5211. How many four-digit "interesting" numbers are there?*

> Answer: 4050
> **Solution**: Let $\overline{abcd}$ be an "interesting" four-digit number. Note that altogether there are 90 two-digit numbers, so that 45 of them are even and 45 of them are odd. Thus, there are 45 different values so that two-digit number $\overline{ab}$ is even and 45 different values so that two-digit number $\overline{cd}$ is odd. We get $45 \cdot 45 = 2025$ possibilities. In a similar way, there are 45 different values so that two-digit number $\overline{ab}$ is odd and 45 different values so that two-digit number $\overline{cd}$ is even. We get $45 \cdot 45 = 2025$ possibilities. So, we get that there are $2025 + 2025 = 4050$ four-digit "interesting" numbers.

Problem 4.11: *A four-digit number is called "ordinary", if it has at least two neighbor digits of different parities, for example 5200, 5211. How many four-digit "ordinary" numbers are there?*

> Answer: 7875
> **Solution**: Note that if a four-digit number $\overline{abcd}$ is not "ordinary", then all its digits are either even or odd. The number of four-digit numbers where all digits are even is $4 \cdot 5 \cdot 5 \cdot 5 = 500$. The number of four-digit numbers where all digits are odd is $5 \cdot 5 \cdot 5 \cdot 5 = 625$. There are 9000 four-digit numbers (from 1000 to 9999). So, the number of four-digit "ordinary" numbers is $9000 - 500 - 625 = 7875$.

Problem 4.12: *A five-digit number divisible by 11 is called "beautiful", if its digits are different from each other and are 1, 2, 3, 4, 8 (in some order). Find the number of all five-digit "beautiful" numbers.*

> Answer: 12
>
> **Solution**: Assume the number $\overline{abcde}$ is "beautiful". We have that
>
> $$\overline{abcde} = \overline{a0000} + \overline{b000} + \overline{c00} + \overline{d0} + e =$$
>
> $$= 9999a + 1001b + 99c + 11d + (a - b + c - d + e). \qquad (4.1)$$
>
> Note that each of the numbers $\overline{abcde}, 9999, 1001, 99, 11$ is divisible by 11, thus from equation 4.1 we get that
>
> $$(a - b + c - d + e) \vdots 11. \qquad (4.2)$$
>
> Note that $a - b + c - d + e$ is an even number, as a, b, c, d, e are $1, 2, 3, 4, 8$. From this and from equation 4.2, it follows that $a - b + c - d + e = 0$. On the other hand $a + b + c + d + e = 1 + 2 + 3 + 4 + 8 = 18$.
>
> It follows that $a + c + e = b + d = 9$, thus a, c, e are the numbers $2, 3, 4$ and b, d are the numbers $1, 8$.
>
> Thus the number of "beautiful" numbers is $3 \cdot 2 \cdot 1 \cdot 2 = 12$.

Problem 4.13: *What is the greatest number of three-digit numbers so that all of them have the same sum of the digits? For example there are* **three** *three-digit numbers so that the sum of the digits of each of them is equal to 2 (that is 101, 110, 200) and there are* **five** *three-digit numbers so that the sum of the digits of each of them is equal to 3 (that is 102, 111, 120, 201, 210).* **Five** *is greater than* **three**, *but five is not the greatest possible answer. So, what is the greatest possible answer?*

> **Answer**: 70
> **Solution**: Below we have a table showing the number of two-digit numbers where the sum of the digits is the same.
>
> | The sum of digits | 1 | 2 | 3 | 4 | 5 | 6 | 7 | 8 | 9 | 10 | 11 | 12 | 13 | 14 | 15 | 16 | 17 | 18 |
> |---|---|---|---|---|---|---|---|---|---|---|---|---|---|---|---|---|---|---|
> | The number | 1 | 2 | 3 | 4 | 5 | 6 | 7 | 8 | 9 | 9 | 8 | 7 | 6 | 5 | 4 | 3 | 2 | 1 |
>
> Assume the sum of the digits of $\overline{abc}$ is n. We have $a + b + c = n$. Note that c can have at most 10 consecutive values. In that case $a + b$ will have at most 10 consecutive values from the table. The number of $\overline{abc}$'s will at most be the sum of 10 consecutive numbers written in the table above. The sum is maximized in the following case:
>
> $$5 + 6 + 7 + 8 + 9 + 9 + 8 + 7 + 6 + 5 = 70 (n = 14)$$

Problem 4.14: *Suppose for numbers a, b, c, d we have $\overline{ab}, \overline{cd}, \overline{ac}, \overline{bd}$ and $\overline{ab} + \overline{cd} = \overline{ac} + \overline{bd}$. Find the number of all possible (a, b, c, d).*

> **Answer**: 810
> **Solution**: We have $10a + b + 10c + d = 10a + c + 10b + d$, thus $c = b$. According to the *multiplication principle* the number of four-tuples (A, b, c, d) is
>
> $$9 \cdot 9 \cdot 10 = 810$$

Problem 4.15: *In how many ways is it possible to put five candies in three pots, if we know that in the first pot is possible to put not more than one candy, in the second pot not more than two candies and in the third pot not more than three candies?*

> Answer: 60
> **Solution**: We have $1+2+3=6$, thus we may have $0,2,3$, or $1,1,3$ or $1,2,2$ candies in the I, II, III pots respectively. The number of possibilities of the arrangements are $\frac{5 \cdot 4}{2}, 5 \cdot 4$ and $5 \cdot \frac{4 \cdot 3}{2}$.
> According to the *addition principle*, the overall number of possibilities is $10 + 20 + 30 = 60$.

Problem 4.16: *In how many ways is it possible to put different books on a bookshelf if the first, second and third books should not be next to the fourth book?*

> Answer: 144
> **Solution**: When the fourth volume is either the first or the last, it is possible to arrange the remaining volumes in $2 \cdot 2 \cdot 4 \cdot 3 \cdot 2 \cdot 1$ ways.
> Otherwise, the number of possibilities is $4 \cdot 2 \cdot 1 \cdot 3 \cdot 2 \cdot 1$.
> According to the *addition principle*, the number of all possibilities is $96 + 48 = 144$.

Problem 4.17: *Four points were chosen on each edge of a triangular pyramid. Consider those 24 points and four vertices of the pyramid. Find the number of lines that pass through two of the considered 28 points.*

> Answer: 294
>
> **Solution**: Consider the following cases for a line a that passes through 2 of the 28 points.
>
> 1. a passes through the points that lay on the same edge of the pyramid. The number of such lines a is 6.
>
> 2. a passes through two points that belong to the same face, which does not include the side. The number of such lines is $4(3 \cdot 4 + 3 \cdot 16) = 240$.
>
> 3. a passes through the points that belong to not intersecting faces. The number of such lines is $3 \cdot 4 \cdot 4 = 48$.
>
> Thus the overall number of possible lines a is $6 + 240 + 48 = 294$.

Problem 4.18: *The squares of a 3×3 board are colored in red, blue or yellow. Any two squares that share a side are colored in different colors. In how many such ways is it possible to color this board? (N.M. Sedrakyan)*

> Answer: 246
>
> **Solution**: Suppose the middle square is red. The four squares that share a side with it are either yellow or blue. Considering all the possibilities to color these four squares we get that the number of possible colorings is
>
> $$2 \cdot 2 \cdot 2 \cdot 2 + 4 \cdot 2 \cdot 2 \cdot 1 \cdot 1 + 4 \cdot 2 \cdot 2 \cdot 1 \cdot 1 + 2 \cdot 1 \cdot 1 \cdot 1 \cdot 1 + 4 \cdot 2 \cdot 2 \cdot 1 \cdot 1 + 2 \cdot 2 \cdot 2 \cdot 2 = 82.$$
>
> So, the overall number of possibilities is $3 \cdot 82 = 246$.

Problem 4.19: *Find the number of positive divisors of 770^7 so that when divided by 3 the remainder of each of them is 1.*

> **Answer**: 2048
> **Solution**: Let us prime-factorize 770^7. We get $770^7 = 2^7 \cdot 5^7 \cdot 7^7 \cdot 11^7$. So, any divisor of 770^7 has the following form $2^a \cdot 5^b \cdot 7^c \cdot 11^d$, where a, b, c, d are non-negative integers less than or equal to 7. Note that if we look for a divisor of 770^7 so that when divided by 3 its remainder is 1, then $a + b + d$ should be even, as when 7 is divided by 3 the remainder is 1 and when 2, 5, 11 are divided by 3 the remainder is 2. This means that we need to have even number of divisors 2, 5, 11 in order to end up with a remainder of 1. As $0 \leq a, b, c, d \leq 7$ and $a + b + d$ should be even. This means that there are 8 options for a, as a is changing from 0 to 7. There are 8 options for b, 8 options for c and 4 options for d (as once a and b are chosen we need to choose d in such a way that $a + b + d$ is even). So, the total number of all quadruples (a, b, c, d) is $8 \cdot 8 \cdot 8 \cdot 4 = 2048$.

Problem 4.20: *There are six identical (same) mathematics books and four different physics books. In how many ways is it possible to put these books on a bookshelf so that one neighbor of a physics book is a mathematics book and the other one is a physics book?*

> **Answer**: 240
> **Solution**: Arrange the mathematics books on an empty shelf. Then, choose two neighbors and place two physics books in between of them. In a similar way, choose another two neighbors and put two psychics books in between of them. According to the *multiplication principle*, the number of all possible arrangements is $\frac{5 \cdot 4}{2} \cdot 4 \cdot 3 \cdot 2 \cdot 1 = 240$.

4.2 Problems from national competitions

Problem 1: *In the figure below, an ant starts at point X and goes to point Y. Given that the ant can walk only on the sides: right and up. In how many ways can the ant reach point Y?*

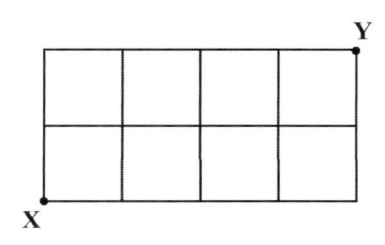

Problem 2: *Find the number of all two-digit numbers, so that for any of them the product of its digits is a composite number (0 is not a composite number).*

Problem 3: *Find the number of all two-digit numbers, so that for any of them its ones digit is divisible by its tens digit.*

Problem 4: *At most, in how many points can four unique straight lines intersect?*

Problem 5: *Find the number of all odd positive divisors of $1 \cdot 2 \cdot 3 \cdot 4 \cdot 5$.*

Problem 6: *If 15 people meet each other and each shakes hands only once with each of the others, how many handshakes will there be?*

Problem 7: *At most, into how many parts can five straight lines divide a plane?*

Problem 8: *How many numbers from 1 to 1000 have exactly three positive divisors?*

Problem 9: *How many seven-digit numbers that are divisible by 6 can be written by using only the digits 0 and 1? Repetition is allowed and both 0 and 1 should be used.*

Problem 10: *How many numbers in the list 1, 2,..., 1000 are either perfect squares or perfect cubes of whole numbers? How many numbers in this list are (at the same time) perfect squares and perfect cubes?*

Problem 11: *Find the number of all positive divisors of 12^{12}.*

Problem 12: *At most, in how many points can four unique circles intersect?*

Problem 13: *How many seven-digit numbers can be written by using the digit 0 four times and the digit 1 three times?*

Problem 14: *How many three-digit numbers can be written such that the digits are in increasing order from left to right and the difference of the last two digits is at least 4?*

Problem 15: *How many four-digit numbers $\overline{abcd}$ can be written by using the digits 1, 2, 3, 4, so that each of these digits is used and $a \neq 1, b \neq 2, c \neq 3, d \neq 4$?*

Problem 16: *In how many ways is it possible to place five different fruits into two different bowls? A bowl can have 0 fruits also.*

Problem 17: *How many even positive whole numbers have exactly five positive divisors?*

Problem 18: *Eight straight lines are drawn in the plane such that no two lines are parallel and no three lines intersect at one point. How many triangles are there so that all sides are on three of these eight lines?*

Problem 19: *How many positive three-digit numbers are there so that for each of them its last digit is equal to the product of its first two digits?*

Problem 20: *How many positive two-digit numbers are there so that for each of them its first digit is greater than its second digit?*

Problem 21: *21 points are in a plane, no three of them lie on a straight line. How many straight lines can be formed using these conditions?*

Problem 22: *Let A be the sum of all digits of a three-digit number. Let B be the sum of all digits of A and let C be the sum of all digits of B. How many three-digit numbers are there so that B and C are different?*

4.2.1 Hints and Solutions

Problem 1: *In the figure below, an ant starts at point X and goes to point Y. Given that the ant can walk only on the sides: either right or up. In how many ways can the ant reach point Y?*

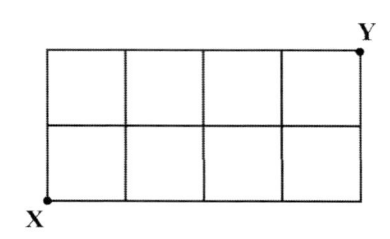

Answer: 15

Hint: Let us label each intersection point with the number representing in how many ways can the ant reach that intersection point (see the picture). Then, the label of Y is 15, as $10 + 5 = 15$.

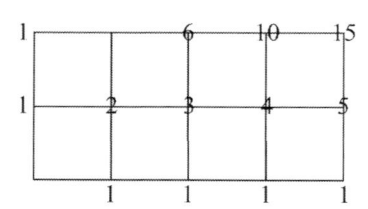

Problem 2: *Find the number of all two-digit numbers, so that for any of them the product of its digits is a composite number (0 is not a composite number).*

Answer: 72

Hint: The product of the digits of a two-digit number is not composite, if either one of its digits is 1 and the other one is a prime number, or if it ends with 0 (as 0 is not a composite number). So, for any of the following two-digit numbers the product of its digits is not a composite number 11, 12, 13, 15, 17, 21, 31, 51, 71, 10, 20, 30, 40, 50, 60, 70, 80, 90. Note that there are 18 numbers written here and there are 90 two-digit numbers in total (from 10 to 99). Then, the answer is $90 - 18 = 72$.

Problem 3: *Find the number of all two-digit numbers, so that for any of them its ones digit is divisible by its tens digit.*

> Answer: 32
> **Hint**: Such numbers are 10, 20, 30, 40, 50, 60, 70, 80, 90, 11, 12, 13, 14, 15, 16, 17, 18, 19, 22, 24, 26, 28, 33, 36, 39, 44, 48, 55, 66, 77, 88, 99.

Problem 4: *At most, in how many points can four unique straight lines intersect?*

> Answer: 6

Problem 5: *Find the number of all odd positive divisors of $1 \cdot 2 \cdot 3 \cdot 4 \cdot 5$.*

> Answer: 4

Problem 6: *If 15 people meet each other and each shakes hands only once with each of the others, how many handshakes will there be?*

> Answer: 105

Problem 7: *At most, into how many parts can five straight lines divide a plane?*

> Answer: 16
> **Hint**: $2 + 2 + 3 + 4 + 5 = 16$

Problem 8: *How many numbers from 1 to 1000 have exactly three positive divisors?*

> Answer: 11
> **Hint**: Those are squares of prime numbers

Problem 9: *How many seven-digit numbers that are divisible by 6 can be written by using only the digits 0 and 1? Repetition is allowed and both 0 and 1 should be used.*

> Answer: 11
> **Hint**: The numbers are 1111110 and $1\star\star\star\star\star 0$, where exactly two $\star$ are equal to 1.

Problem 10: *How many numbers in the list 1, 2,..., 1000 are either perfect squares or perfect cubes of whole numbers? How many numbers in this list are (at the same time) perfect squares and perfect cubes?*

> Answer: 38, 3
> **Hint**: One can easily check that there are 31 perfect squares and 10 perfect cubes. Note that 1, 64, 729 are (at the same time) perfect squares and perfect cubes. So, the answer is 38 (perfect squares or perfect cubes) and 3 (perfect squares and perfect cubes).

Problem 11: *Find the number of all positive divisors of 12^{12}.*

> Answer: 325
> **Hint**: $12^{12} = 2^{24} \cdot 3^{12}$

Problem 12: *At most, in how many points can four unique circles intersect?*

> Answer: 12
> **Hint**: $2 \cdot 6 = 12$

Problem 13: *How many seven-digit numbers can be written by using the digit 0 four times and the digit 1 three times?*

> Answer: 15
> **Hint**: $\frac{6 \cdot 5}{2} = 15$

Problem 14: *How many three-digit numbers can be written such that the digits are in increasing order from left to right and the difference of the last two digits is at least 4?*

> Answer: 20
> **Hint**: $1 \cdot 4 + 2 \cdot 3 + 3 \cdot 2 + 4 \cdot 1 = 20$

Problem 15: *How many four-digit numbers $\overline{abcd}$ can be written by using the digits 1, 2, 3, 4, so that each of these digits is used and $a \neq 1, b \neq 2, c \neq 3, d \neq 4$?*

> Answer: 9
> **Hint**: $3 \cdot 3 \cdot 1 = 9$

Problem 16: *In how many ways is it possible to place five different fruits into two different bowls? A bowl can have 0 fruits also.*

> Answer: 32
> **Hint**: $2 \cdot 2 \cdot 2 \cdot 2 \cdot 2 = 32$.

Problem 17: *How many even positive whole numbers have exactly five positive divisors?*

> Answer: 1

Problem 18: *Eight straight lines are drawn in the plane such that no two lines are parallel and no three lines intersect at one point. How many triangles are there so that all sides are on three of these eight lines?*

> Answer: 55
> **Hint**: $\frac{8 \cdot 7 \cdot 6}{3 \cdot 2 \cdot 1} - 1 = 55$

Problem 19: *How many positive three-digit numbers are there so that for each of them its last digit is equal to the product of its first two digits?*

> Answer: 32
> **Hint**: Let number $\overline{abc}$ be so that $a \cdot b = c$. So, numbers $\overline{ab}$ are:
>
> $$10, 11, \ldots, 19, 20, 21, 22, 23, 24, 30, 31, 32, 33,$$
>
> $$40, 41, 42, 50, 51, 60, 61, 70, 71, 80, 81, 90, 91.$$

Problem 20: *How many positive two-digit numbers are there so that for each of them its first digit is greater than its second digit?*

> Answer: 45
> **Hint**: $\frac{10 \cdot 9}{2} = 45$

Problem 21: *21 points are in a plane, no three of them lie on a straight line. How many straight lines can be formed using these conditions?*

> Answer: 210
> **Hint**: $\frac{21 \cdot 20}{2} = 210$

Problem 22: *Let A be the sum of all digits of a three-digit number. Let B be the sum of all digits of A and let C be the sum of all digits of B. How many three-digit numbers are there so that B and C are different?*

> Answer: 45
> **Hint**: Note that $A \leq 27$. Thus $B = 10, A = 19, 1 + 2 + \cdots + 9 = 45$.

Chapter 5
Counting in two different ways

In this chapter, we provide a family of examples and problems that can be solved by counting the same value (mathematical expression) in **two different ways**.

> **Example 5.1:**
>
> *Nine students went hiking. Each of them brought either 2 or 4 apples, and they kept all their apples in the same basket. At the end of the hike there were no apples left. Is it possible that each student ate exactly three apples?*

> **Example 5.1: Solution**
> **Solution:** Proof by contradiction argument. Assume that each student ate exactly three apples. As there are nine students, then on the one hand the number of eaten apples is equal to $9 \cdot 3 = 27$.
> On the other hand, if m students brought 2 apples, then $9 - m$ students brought 4 apples. So, the overall number of apples is $2m + 4(9 - m)$. We get that $2m + 4(9 - m) = 27$. Thus $m = 4.5$, which is not possible.

Example 5.2:

Some of the squares of a board were colored. We know that seven squares out of nine squares of any 3×3 part of this board are colored. Prove that there is a 2×2 part of this board so that all its squares are colored.

Example 5.2: Proof

Proof: Consider a 6×6 board. Let us use two ways of counting the number n of colored squares on this board. First, divide this 6×6 board into 4 boards of sizes 3×3. According to condition of the problem, we have

$$m \geq 4 \cdot 7 = 28$$

Now, let us divide the 6×6 board into nine boards of size 2×2. If there is at least one non-colored square in these nine boards, then we have

$$n \leq 3 \cdot 9 = 27$$

Note that these two equations cannot hold true at the same time. This ends the proof.

5.1 Practice Problems

Problem 5.1. *A soccer ball is made of 32 parts: white hexagons and black pentagons, so that each pentagon shares all its five sides with five hexagons and each hexagon shares three sides with three pentagons and three sides with three other hexagons (see the figure). How many hexagons are there?*

Problem 5.2. *Given $n + m$ numbers on a circle, n ones and m negative ones. Let p be the number of neighboring ones and let q be the number of neighboring negative ones. Prove that $n - m = p - q$.*

Problem 5.3. *There are less than 32 students in a class. Each student is friends with exactly five girls and exactly six boys in that class. How many girls are there in that class?*

Problem 5.4. *Provide an example of a class described in problem 5.3 with 22 students in the class.*

Problem 5.5. *There are nine rooks on a 9×9 chessboard, so that they do not hit each other. Move each rook to one of the neighbor squares. Prove that, after this move, there are two rooks that hit each other.*

Problem 5.6. *In each square of a 4×4 board is written 1 or 0, so that the sum of the numbers in each 2×2 board is the same and the sum of the numbers in each 3×3 board is the same. Can we state that all the numbers in this 4×4 board are the same?*

Problem 5.7. *Let triangle ABC be divided into 2022 triangles so that each small triangle has an angle bigger than $120°$. Prove that it is possible to divide triangle ABC into 2021 triangles so that each of them has an angle bigger than $120°$. (N.M.Sedrakyan)*

5.1.1 Solutions of practice problems

Problem 5.1: *A soccer ball is made of 32 parts: white hexagons and black pentagons, so that each pentagon shares all its five sides with five hexagons and each hexagon shares three sides with three pentagons and three sides with three other hexagons (see the figure). How many hexagons are there?*

Answer. 20
Solution: Let m be the number of hexagons, then the number of pentagons is $32 - m$. Let us count in two different ways the number of all shared sides that are between parts of different colors.
On the one hand, this number is $5(32 - m)$.
On the other hand, this number is $3m$.
So, $3m = 5(32 - m)$. We get $8m = 5 \cdot 32$ and $m = 20$.

Problem 5.2: *Given $n + m$ numbers on a circle, n ones and m negative ones. Let p be the number of neighboring ones and let q be the number of neighboring negative ones. Prove that $n - m = p - q$.*

Proof: For each pair of neighboring numbers let us write their sum between them. Note that, the sum of the newly written numbers is twice as much as the sum of the initial numbers. That is $2(n \cdot 1 + m \cdot (-1)) = 2(n - m)$. On the other hand, there are p twos, q negative twos and all other numbers are zeros.
Thus, it follows that $2p + (-2)q = 2(p - q)$.
So, $2(n - m) = 2(p - q)$. Therefore $n - m = p - q$.

Problem 5.3: *There are less than 32 students in a class. Each student is friends with exactly five girls and exactly six boys in that class. How many girls are there in that class?*

> Answer. 10
> **Solution**: Let m be the number of boys in that classroom and n be the number of girls. Let us count the number of boy-girl friends.
> As each student needs to be friends with exactly five girls and there are m boys, then the number of girls is $5m$. On the other hand, as each student needs to be friends with exactly six boys and there are n girls, then the number of girls is $6n$.
> So, $5m = 6n$. We get $m = 6k, n = 5k$, where k is a natural number. According to the condition of the problem, there are less than 32 students and each student is friends with exactly 11 students. So, the number of students is not less than 12. Thus, we get $12 \leq 11k < 32$. Therefore $k = 2$. Then $n = 10$.

Problem 5.4: *Provide an example of a class described in problem 5.3 with 22 students in the class.*

> Answer. 10 girls and 12 boys
> **Solution**: There should be 10 girls and 12 boys in the classroom. Divide the girls into groups A and B, so that in each group there are 5 girls. Similarly, divide the boys into groups C and D, so that in each group there are 6 boys.
> Let us use $X - Y$ notation to describe that each student in the group X is friends with everyone in the group Y.
> Note that, the following examples satisfy the conditions of the problem $A - B, A - C, B - D, C - D$.

Problem 5.5: *There are nine rooks on a 9×9 chessboard, so that they do not hit each other. Move each rook to one of the neighbor squares. Prove that, after this move, there are two rooks that hit each other.*

> **Proof**: Proof by contradiction argument. Assume that after moving these nine rooks there are no rooks that hit each other.
>
> Number the columns of this chessboard (from left to right) and number its rows (from bottom to top) by numbers 1 to 9.
>
> For each rook consider the sum of the column number and the row number. Let us call it its "age".
>
> After moving all nine rooks, compute the sum of all their "ages". According to the condition of the problem, it is $2(1 + 2 + \cdots + 9)$.
>
> Note also that, before moving the rooks this sum was $2(1 + 2 + \cdots + 9)$.
>
> After the move, the "age" of each rook is greater by 1 than its initial "age". So, after the move, the sum of the "ages" is different (by an odd number) from $2(1 + 2 + \cdots + 9)$. This is not possible and is a contradiction.
>
> So, after the move, there are two rooks that hit each other.
> This ends the proof.

Problem 5.6: *In each square of a 4 × 4 board is written 1 or 0, so that the sum of the numbers in each 2 × 2 board is the same and the sum of the numbers in each 3 × 3 board is the same. Can we state that all the numbers in this 4 × 4 board are the same?*

Answer. Yes.

Solution: Let m be the sum of all numbers in 2 × 2 board and let n be the sum of all numbers 3 × 3 board.

Consider all 2 × 2 boards in 4 × 4 boards. The number of such boards is $3 \cdot 3 = 9$. In each square of 4 × 4 board write the number of 2 × 2 boards (see the picture).

1	2	2	1
2	4	4	2
2	4	4	2
1	2	2	1

Now, consider all 3 × 3 boards in 4 × 4 board. The number of such boards is $2 \cdot 2 = 4$. In each square of 4 × 4 board write the number of 3 × 3 boards (see the picture).

1	2	2	1
2	4	4	2
2	4	4	2
1	2	2	1

Note that these two boards are the same. So

$$9m = 4n$$

As $0 \leq m \leq 4$ and $0 \leq n \leq 9$, then $m = n = 0$ or $m = 4, n = 9$. So, all the elements in 4 × 4 board are the same (0 or 1).

Problem 5.7: *Let triangle ABC be divided into 2022 triangles so that each small triangle has an angle bigger than 120°. Prove that it is possible to divide triangle ABC into 2021 triangles so that each of them has an angle bigger than 120°. (N.M.Sedrakyan)*

Proof: Consider all vertices of these 2022 triangles. Note that three of them are A, B, C. Suppose m of these vertices are either inside triangle ABC or are inside another triangle. Denote the number of all other vertices by n.

Let us count in two different ways the sum of all the angles of these 2022 triangles.

On the one hand, this sum is $2022 \cdot 180°$. On the other hand, this sum is $180° + 180°m + 360°n$.

So $1 + m + 2n = 2022$. Thus $m + 2n = 2021$.

Note that, if one of the angles in triangle ABC is not greater than 120°, then the number of all angles greater than 120°: 2022 is not more than $(m + 2n)$. That is $m + 2n \geq 2022$. We get that $2021 \geq 2022$, which is impossible.

We get that one of the angles of triangle ABC is greater than 120°. So, it is possible to divide it into 2021 triangles (see the picture).

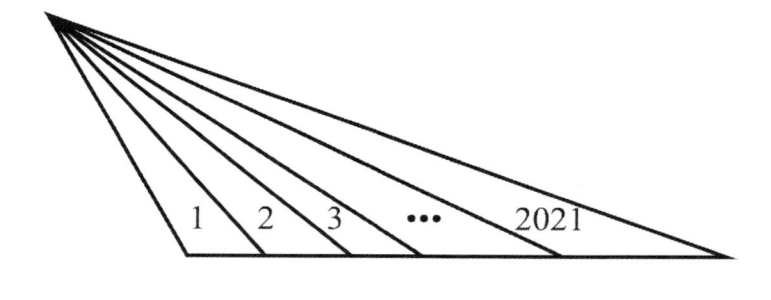

Chapter 6

Triangle inequality

The following inequality is called the **triangle inequality** and is one of the most important inequalities in geometry.

For any points A, B and C we have $AB + BC \geq AC$.

$AB + BC = AC$ if and only if B is on line AC. Otherwise $AB + BC > AC$.

> **Example 6.1:**
>
> *The two sides of isosceles triangle have lengths 4 and 9. Find the length of the base of the triangle.*

> **Example 6.1: Solution**
> **Solution:** If the length of the base of the triangle is 9, then the lengths of the sides should be 4. However, $4 + 4 < 9$ contradicts the triangle inequality. Note that there exists a triangle with sides $9, 9, 4$, since $9+4 > 9$ and $9 + 9 > 4$. Thus the length of the base is 4.

> **Example 6.2:**
>
> *Prove that each side of any triangle is bigger than half of its perimeter.*

> **Example 6.2: Proof**
> **Proof:** Suppose the sides are a, b, c. According to the triangle inequality $a < b + c$, thus $2a < a + b + c$, and $a < \frac{a+b+c}{2}$. The statement is proven.

6.1 Practice Problems

Problem 6.1. *For any points A, B, C prove that $|AB - BC| \leq AC$.*

Problem 6.2. *Let ABC be a triangle, so that $AC = 3.2, BC = 0.7$. Find the length of side AB, if it is a natural number.*

Problem 6.3. *Prove that $2ab + 2bc + 2ac > a^2 + b^2 + c^2$, where a, b, c are side lengths of a triangle.*

Problem 6.4. *For any points A, B, C, D prove that $AB + BC + CD \geq AD$.*

Problem 6.5. *The distance between cities A and B is 122 km, the distance between cities B and C is 78km, the distance between cities C and D is 110 km, the distance between cities A and D is 310km. Find the sum of the distances AC and BD.*

Problem 6.6. *For quadrilateral $ABCD$ we know that $AB = CD$. Let M be any point, prove that $MA + MB + MC \geq MD$.*

Problem 6.7. *Let $ABCD$ be a convex quadrilateral. Find a point M so that $MA + MB + MC + MD$ is the smallest possible.*

Problem 6.8. *Let M and N be two points inside triangle ABC. Prove that MN is less than the length of the largest side of ABC.*

Problem 6.9. *Let $ABCD$ be a convex quadrilateral. Prove that $AC + BD > AB + CD$.*

Problem 6.10. *Let M be a point inside triangle ABC. Prove that $AB + AC > MB + MC$.*

Problem 6.11. *Let M and N be given points on sides AB and BC of triangle ABC, respectively. Let K be the midpoint of line segment MN. Prove that $AK + CK > AM + CN$.*

Problem 6.12. *Let M and N be given points inside triangle ABC. Prove that*

$$MA + MB + MC < NA + NB + NC + MN.$$

Problem 6.13. *Let M be a point inside triangle ABC. Prove that*

$$MA + MB + MC + min(MA, MB, MC) < AB + BC + AC.$$

Here $min(a, b, c)$ is the smallest number among a, b, c. (N.M. Sedrakyan)

Problem 6.14. *Let M and N be the midpoints of sides AB and CD of quadrilateral $ABCD$, respectively. Prove that $MN \leq \frac{BC+AD}{2}$.*

Problem 6.15. *Let M, N, P, K be points given on sides AB, BC, CD, DA of quadrilateral $ABCD$, respectively. Prove that $MN + NP + PK + MK \geq 2AC$.*

Problem 6.16. *Let a, b, c be side lengths of a triangle. Prove that $\frac{a}{b+c-a} + \frac{b}{c+a-b} + \frac{c}{a+b-c} \geq 3$.*

Problem 6.17. *Let a, b, c be side lengths of a triangle. Prove that there exist numbers m, n, k, so that $a = n + k, b = m + k, c = m + n$.*

Problem 6.18. *Let a, b, c be side lengths of a triangle. Prove that $a^2b(a-b) + b^2c(b-c) + c^2a(c-a) \geq 0$.*

6.1.1 Solutions of practice problems

Problem 6.1: *For any points A, B, C prove that $|AB - BC| \leq AC$.*

> **Solution**: Without loss of generality we may assume that $AB \geq BC$.
> We have that $|AB - BC| = AB - BC$.
> According to the triangle inequality $AB \leq BC + AC$, therefore $AB - BC \leq AC$, thus $|AB - BC| \leq AC$.

Problem 6.2: *Let ABC be a triangle, so that $AC = 3.2, BC = 0.7$. Find the length of side AB, if it is a natural number.*

> Answer. 3
> **Solution**: According to the triangle inequality $AB + BC > AC$ and $AB < BC + AC$, thus $AB > 2.5$ and $AB < 3.9$.
> We have that the length of AB is a natural number. Thus $AB = 3$.

Problem 6.3: *Prove that $2ab + 2bc + 2ac > a^2 + b^2 + c^2$, where a, b, c are side lengths of a triangle.*

> **Solution**: Let us form the difference of the left and right parts of the inequality, and transform it.
>
> $$2ab + 2bc + 2ac - (a^2 + b^2 + c^2) = (ab + ac - a^2) + (ab + bc - b^2) + (ac + bc - c^2) =$$
>
> $$= a(b + c - a) + b(a + c - b) + c(a + b - c).$$
>
> We know that $b + c - a > 0, a + c - b > 0, a + b - c > 0$. Thus, it follows that $2ab + 2bc + 2ac - (a^2 + b^2 + c^2) > 0$, thus $2ab + 2bc + 2ac > a^2 + b^2 + c^2$.

Problem 6.4: *For any points A, B, C, D prove that $AB+BC+CD \geq AD$.*

> **Solution**: According to the triangle inequality, we have $AB + BC \geq AC$ and $AC + CD \geq AD$. Thus
>
> $$AB + BC + CD - AD = (AB + BC - AC) + (AC + CD - AD) \geq 0.$$
>
> Therefore
> $$AB + BC + CD \geq AD.$$

Problem 6.5: *The distance between cities A and B is 122 km, the distance between cities B and C is 78km, the distance between cities C and D is 110 km, the distance between cities A and D is 310km. Find the sum of the distances AC and BD.*

> Answer. 388 km
> **Solution**: We know that $AB = 122, BC = 78, CD = 110, AD = 310$. Note that $AB + BC + CD - AD = 122 + 78 + 110 - 310 = 0$. Moreover, $AB + BC + CD - AD \geq 0$ according to problem 6.4.
> Furthermore, from the problem's statement we have that $AB+BC+CD-AD = 0$, when $AB + BC = AC$ and $AC + CD = AD$.
> That is, point B belongs to the AC segment, and point C belongs to the BD segment (see the picture).
>
> A———B—C———D
>
> We can conclude from the statement that the points A, B, C and D belong to the same line. In that case:
>
> $$AC + BD = (AB + BC) + (BC + CD) = 122 + 2 \cdot 78 + 110 = 388.$$

Problem 6.6: *For quadrilateral $ABCD$ we know that $AB = CD$. Let M be any point, prove that*

$$MA + MB + MC \geq MD.$$

Proof: According to the triangle inequality, for the three-tuples M, A, B and M, C, D we have $MA + MB \geq AB$ and $MC + CD \geq MD$. Summing the two inequalities and considering that $AB = CD$, we get that $MA + MB + MC \geq MD$.

Problem 6.7: *Let $ABCD$ be a convex quadrilateral. Find a point M so that $MA + MB + MC + MD$ is the smallest possible.*

Solution: According to the triangle inequality, for the three-tuples M, A, C and M, B, D we have $MA + MC \geq AC$ and $MB + MD \geq BD$. Summing the two inequalities we get

$$MA + MC + MB + MD \geq AC + BD.$$

The equality will hold if and only if $MA + MC = AC$ and $MB + MD = BD$.
That is when M belongs to AC and BD at the same time.
Hence, the sum is minimized when M is the intersection point of two diagonals of $ABCD$.

Problem 6.8: *Let M and N be two points inside triangle ABC. Prove that MN is less than the length of the largest side of ABC.*

Proof: Draw two lines passing trough M, N and perpendicular to MN (see the picture).

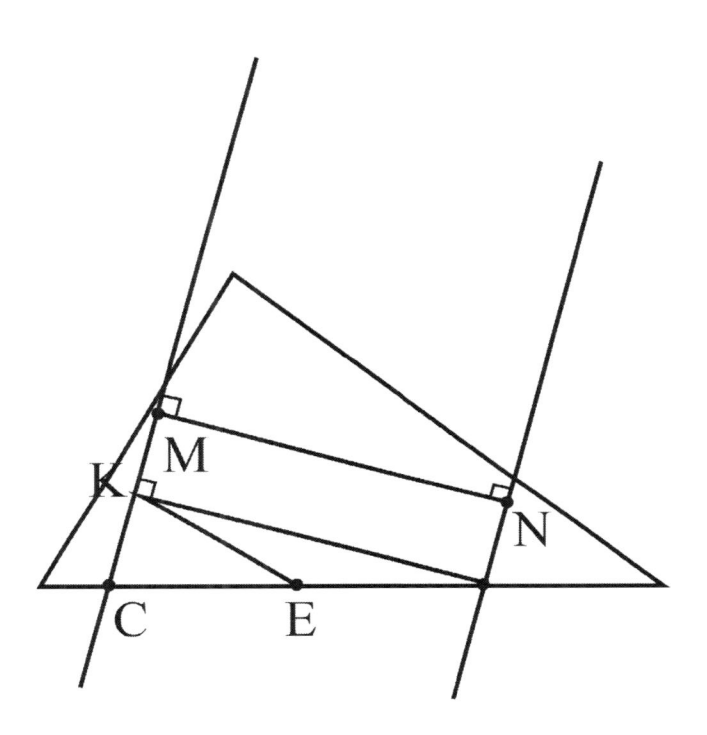

Each of those lines crosses at least two sides of the triangle.

According to *pigeonhole principle*, there exists a side so that these two lines have common points on that side. Let these points be C and D. We have that $DK \parallel MN$.

Suppose E is the midpoint of CD. Hence, $KE = CE$ and $KD = MN$. According to triangle inequality, $KE + ED \geq KD$, therefore $CE + ED \geq CD$, thus $CD \geq MN$.

On the other hand CD is smaller than the biggest side of the triangle, hence MN is also smaller from the biggest side.

Problem 6.9: *Let $ABCD$ be a convex quadrilateral. Prove that*
$$AC + BD > AB + CD.$$

Proof: Suppose the diagonals of quadrilateral $ABCD$ intersect at point M (see the picture).

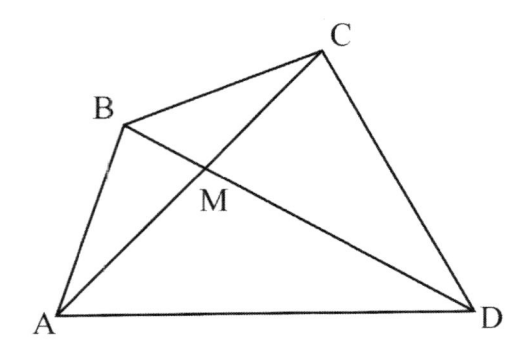

Consider triangles MAB and MCD. According to the triangle inequality we have that
$$MA + MB > AB,$$
$$MC + MD > CD.$$
Summing up these two inequalities, we get that
$$(MA + MB) + (MC + MD) = (MA + MC) + (MB + MD) = AC + BD.$$
Thus, it follows that
$$AC + BD > AB + CD.$$

Problem 6.10: *The point M lies in the ABC triangle. Prove that*
$$AB + AC > MB + MC.$$

Proof: Let N be the intersection point of the extension of line segment BM and side AC (see the picture).

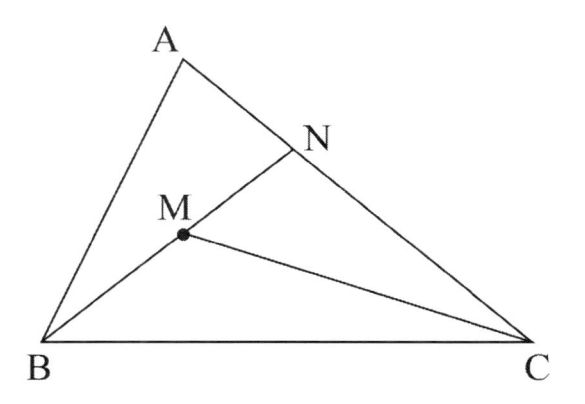

According to the triangle inequality, we have that
$$AB + AN > BN = MB + MN,$$
$$MN + NC > MN.$$
Summing up these two inequalities, we get that
$$AB + AN + NC > MB + MC,$$
$$AB + AC > MB + MC.$$

Problem 6.11: *Let M and N be given points on sides AB and BC of triangle ABC, respectively. Let K be the midpoint of line segment MN. Prove that*

$$AK + CK > AM + CN.$$

Proof: Let $CK = KE$ (see the picture).

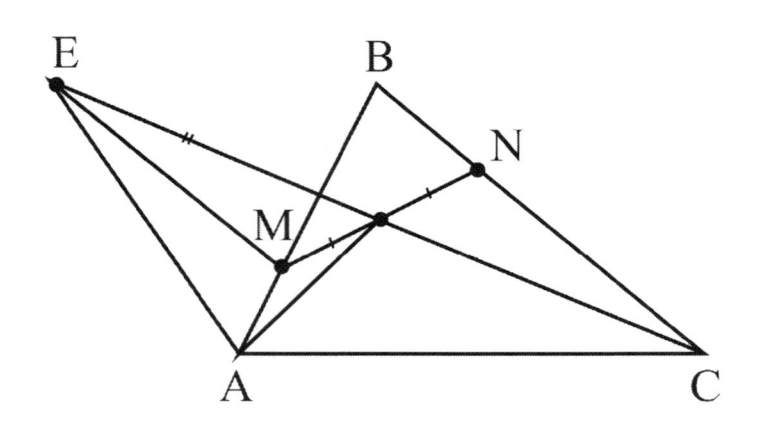

We have $KM = KN, KE = CK$ and $\angle MKE = \angle NKC$.
We get that triangles MEK and NCK are congruent.
So, $ME = CN$ and $\angle MEK = \angle NCK$.
Thus, it follows that $ME \parallel BC$.
This means that M is in triangle AEK.
According to problem 10 we have

$$AK + EK > AM + EM.$$

Thus, it follows that

$$AK + CK > AM + CN.$$

Problem 6.12: *Let M and N be given points inside triangle ABC. Prove that*
$$MA + MB + MC < NA + NB + NC + MN.$$

Proof: Point M belongs to one of the triangles NAB, NBC, NAC. Assume that M belongs to triangle NAB (see the picture).

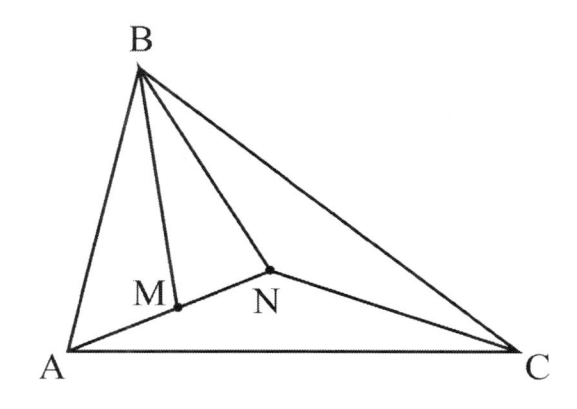

According to the triangle inequality and problem 10, we have

$$MA + MB < NA + NB$$

According to the triangle inequality, we also have that

$$MC \leq MN + NC.$$

Summing up these two inequalities, we get that

$$MA + MB + MC < NA + NB + NC + MN.$$

Problem 6.13: *Let M be a point inside triangle ABC. Prove that*

$$MA + MB + MC + min(MA, MB, MC) < AB + BC + AC.$$

Here $min(a, b, c)$ is the smallest number among a, b, c. (N.M. Sedrakyan)

Proof: Let C_1, A_1, B_1 be the midpoints of sides C_1, A_1, B_1, respectively. Note that any point that is inside triangle ABC belongs to at least two of these three quadrilaterals AC_1A_1C, AB_1A_1B, BC_1B_1C.
Suppose M belongs to AC_1A_1C, AB_1A_1B (see the picture).

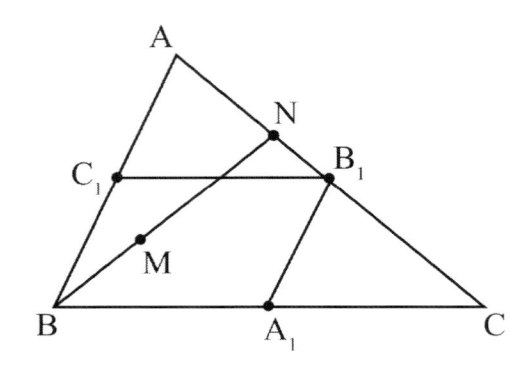

Repeating the steps of problem 10 and using problem 4, we get

$$MA + MB < BA_1 + A_1B_1 + AB_1.$$

On the other hand, A_1B_1 is a midsegment and is equal to $\frac{AB}{2}$. Thus

$$MA + MB < \frac{BC + AB + AC}{2}.$$

In a similar way, one can get that $MA + MC < \frac{AB+BC+AC}{2}$.
Summing up last two inequalities, we get that

$$2MA + MB + MC < AB + BC + AC.$$

Moreover $min(MA, MB, MC) \leq MA$.
Summing up last two inequalities, we get that

$$MA + MB + MC + min(MA, MB, MC) < AB + BC + AC.$$

Problem 6.14: *Let M and N be the midpoints of sides AB and CD of quadrilateral $ABCD$, respectively. Prove that*

$$MN \leq \frac{BC + AD}{2}.$$

Proof: Let K be the midpoint of diagonal AC (see the picture).

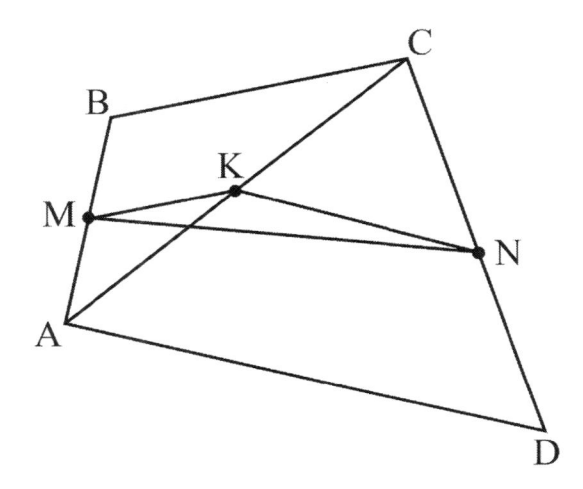

Note that MK and KN are midsegments in triangles ABC and ACD, respectively. According to the midsegment property, we have that

$$MK = \frac{BC}{2}, KN = \frac{AD}{2}.$$

According to the triangle inequality, we have that

$$MK + KN \geq MN$$

Thus, it follows that

$$\frac{BC + AD}{2} \geq MN.$$

Problem 6.15: *Let M, N, P, K be points on sides AB, BC, CD, DA of quadrilateral $ABCD$, respectively. Prove that*

$$MN + NP + PK + MK \geq 2AC.$$

Proof: Let B_1, C_1, D_1, A_1 be the midpoints of line segments MN, NP, PK, MK, respectively (see the figure).

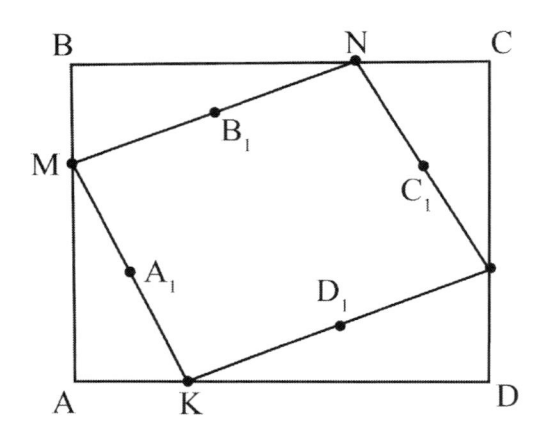

According to problem 4, we have

$$AA_1 + A_1C_1 + C_1C \geq AC.$$

Note that

$$AA_1 = \frac{MK}{2}, CC_1 = \frac{NP}{2}.$$

Thus, it follows that

$$\frac{MK}{2} + A_1C_1 + \frac{NP}{2} \geq AC.$$

According to problem 14, we have that

$$\frac{MN + PK}{2} \geq A_1C_1.$$

Adding the last two inequalities, we get that

$$MN + NP + PK + MK \geq 2AC.$$

Problem 6.16: *Let a, b, c be side lengths of a triangle. Prove that*

$$\frac{a}{b+c-a} + \frac{b}{c+a-b} + \frac{c}{a+b-c} \geq 3.$$

Proof: Let $b + c - a = 2m, c + a - b = 2n, a + b - c = 2k$.
According to the triangle inequality, we have $m > 0, n > 0, k > 0$.
Summing the first two equations, we get that $c = m + n$.
In a similar way, one can get that $a = n + k$ and $b = m + k$.
The inequality that we need to prove has the following form

$$\frac{n+k}{2m} + \frac{m+k}{2n} + \frac{m+n}{2k} \geq 3.$$

Writing in the common denominator form and cross-multiplying, we get that

$$n^2k + k^2n + m^2k + k^2m + m^2n + n^2m \geq 6mnk.$$

Note that, the last inequality may be rewritten in the following form

$$k(m-n)^2 + m(n-k)^2 + n(m-k)^2 \geq 0.$$

The last inequality obviously holds true, as the left hand-side is a sum of non-negative expressions.

Problem 6.17: *Let a, b, c be side lengths of a triangle. Prove that there exist numbers m, n, k, so that $a = n + k, b = m + k, c = m + n$.*

Proof: See the proof of problem 16.

Problem 6.18: *Let a, b, c be side lengths of a triangle. Prove that*

$$a^2b(a-b) + b^2c(b-c) + c^2a(c-a) \geq 0.$$

Proof: Let
$$a = n+k, b = m+k, c = m+n,$$
where $m > 0, n > 0, k > 0$.

Then, we need to prove the following inequality
$$A = (n+k)^2(m+k)(n-m) + (m+k)^2(m+n)(k-n) + (m+n)^2(n+k)(m-k) \geq 0$$

We have
$$A = ((m+n)^2(n+k) - (n+k)^2(m+k))m + ((n+k)^2(m+k) - (m+k)^2(m+n))n +$$
$$+((m+k)^2(m+n) - (m+n)^2(n+k))k = ((m+n)^2 - (n+k)(m+k))(mn+mk) +$$
$$+((n+k)^2 - (m+k)(m+n))(mn+kn) + ((m+k)^2 - (m+n)(n+k))(mk+kn) =$$
$$= mn((m+n)^2 - (n+k)(m+k) + (n+k)^2(m+k)(m+n)) + mk((m+n)^2 -$$
$$-(n+k)(m+n) + (m+k)^2 - (m+n)(n+k)) + kn((m+k)^2 - (m+k)(m+n) +$$
$$+(n+k)^2 - (m+n)(n+k)) = mn(2n^2 - 2nk) + mk(2m^2 - 2nk) +$$
$$+kn(2k^2 - 2mn) = 2mn(k-n)^2 + 2mk(m-n)^2 + 2kn(k-m)^2 \geq 0.$$

Thus, it follows that
$$A \geq 0.$$

6.2 Problems from national competitions

Problem 1: *Let point C be the midpoint of line segment AB and point D be the midpoint of line segment BC. The distance between the midpoints of line segments AC and CD is 12 cm. Find the length of line segment AB.*

Problem 2: *Let OD be a ray inside angle BOC (which is adjacent to angle AOC), so that $\angle COD = 42°$. Find the angle between the angle bisectors of AOC and BOD.*

Problem 3: *Points A, B, C, D are on one line. Given that $AB = 6$, $BC = 3$, $CD = 12$, $AD = 15$. Find the length of the longest segment between any two of these points.*

Problem 4: *Let BD be a median of triangle ABC. Given that the perimeter of triangle ABC is 32cm and the perimeter of triangle ABD is 24cm. Find the length of BD.*

Problem 5: *Given a square $ABCD$ of side length 10. Let M be a point on side BC, so that $BM : MC = 4 : 6$ and let N be a point on side CD, so that $CN : ND = 8 : 2$. Find the area of triangle AMN.*

Problem 6: *Let D be a given point on the extension of side AB of triangle ABC. Given that the angle bisector of angle DBC is parallel to AC. Given also that $AB = 2AC$ and the perimeter of ABC is 35cm. Find the smallest side of triangle ABC.*

Problem 7: *Let M be a point on side BC of rectangle $ABCD$. Given that the area of rectangle $ABCD$ is 38 cm^2. Find the area of triangle AMD.*

Problem 8: *Let several lines on the plane create with each other angles of sizes $10°, 20°, 30°, 40°, 50°, 60°, 70°, 80°, 90°$. What is the smallest number of such lines?*

Problem 9: *Consider the angle bisector of angle B of triangle ABC. Two lines passing through points A and C, which are perpendicular to this angle bisector intersect lines BC and AB at points K and M, respectively. Given that $BM = 8, CK = 1$. Find the sum of all possible values of line segment AB.*

Problem 10: *Let E and D be points on sides AB and BC of triangle ABC, so that $\angle BEC = \angle BDA$ and $BD = BE$. Given that points B and M are on two opposite sides of line AC and $\angle BAC = \angle CAM$. Prove that $BC \parallel AM$.*

Problem 11: *Consider four points which are on one line and one more point that is not on that line. Consider six triangles that are formed by these points. At most, how many of these triangles can be isosceles?*

Problem 12: *Let D be a point on hypotenuse AB of right triangle ABC and let K be a point on leg AC. Given that $BD = AC$ and $KC = AD$. Prove that $\angle BAC = 2\angle CBK$.*

6.2.1 Hints and Solutions

Problem 1: *Let point C be the midpoint of line segment AB and point D be the midpoint of line segment BC. The distance between the midpoints of line segments AC and CD is 12 cm. Find the length of line segment AB.*

> Answer: 32 cm
> **Hint**: If $CD = 2a$, then $3a = 12$ and $AB = 8a$.

Problem 2: *Let OD be a ray inside angle BOC (which is adjacent to angle AOC), so that $\angle COD = 42°$. Find the angle between the angle bisectors of AOC and BOD.*

> Answer: 111°
> **Hint**: We have that $\angle AOC + \angle BOD = 180° - 42° = 138°$ and
> $$42° + \frac{138°}{2} = 111°.$$

Problem 3: *Points A, B, C, D are on one line. Given that $AB = 6$, $BC = 3$, $CD = 12$, $AD = 15$. Find the length of the longest segment between any two of these points.*

> Answer: 15 cm
> **Hint**: According to triangle inequality:
> $$AC \leq AB + BC = 9, BD \leq BC + CD = 15.$$

Problem 4: *Let BD be a median of triangle ABC. Given that the perimeter of triangle ABC is 32cm and the perimeter of triangle ABD is 24cm. Find the length of BD.*

> Answer: 8 cm
> **Hint**: $2BD = 2 \cdot 24 - 32$.

Problem 5: *Given a square ABCD of side length 10. Let M be a point on side BC, so that BM : MC = 4 : 6 and let N be a point on side CD, so that CN : ND = 8 : 2. Find the area of triangle AMN.*

> Answer: 46
> **Hint**: $BM = 4, CM = 6, CN = 8$ and $DN = 2$ and
> $$10 \cdot 10 - \frac{10 \cdot 4}{2} - \frac{6 \cdot 8}{2} - \frac{10 \cdot 2}{2} = 46.$$

Problem 6: *Let D be a given point on the extension of side AB of triangle ABC. Given that the angle bisector of angle DBC is parallel to AC. Given also that AB = 2AC and the perimeter of ABC is 35 cm. Find the smallest side of triangle ABC.*

> Answer: 7 cm
> **Hint**: $\angle BAC = \frac{1}{2}\angle DBC = \angle ACB$.

Problem 7: *Let M be a point on side BC of rectangle ABCD. Given that the area of rectangle ABCD is 38 cm^2. Find the area of triangle AMD.*

> Answer: $19 cm^2$
> **Hint**: Consider point N so that quadrilateral $ABCD$ is a rectangle.

Problem 8: *Let several lines on the plane create with each other angles of sizes 10°, 20°, 30°, 40°, 50°, 60°, 70°, 80°, 90°. What is the smallest number of such lines?*

> Answer: 5
> **Hint**: If n is the number of lines, then $\frac{n(n-1)}{2}$ is the number of formed pairs.

Problem 9: *Consider the angle bisector of angle B of triangle ABC. Two lines passing through points A and C, which are perpendicular to this angle bisector intersect lines BC and AB at points K and M, respectively. Given that $BM = 8, CK = 1$. Find the sum of all possible values of line segment AB.*

> Answer: 16
> **Hint**: Note that points A and K are symmetric with respect to the angle bisector of angle B. Note also that points C and M are symmetric with respect to the angle bisector of angle B.

Problem 10: *Let E and D be points on sides AB and BC of triangle ABC, so that $\angle BEC = \angle BDA$ and $BD = BE$. Given that points B and M are on two opposite sides of line AC and $\angle BAC = \angle CAM$. Prove that $BC \| AM$.*

> **Hint**: Triangles ABD and CBE are congruent, so $AB = BC$.

Problem 11: *Consider four points which are on one line and one more point that is not on that line. Consider six triangles that are formed by these points. At most, how many of these triangles can be isosceles?*

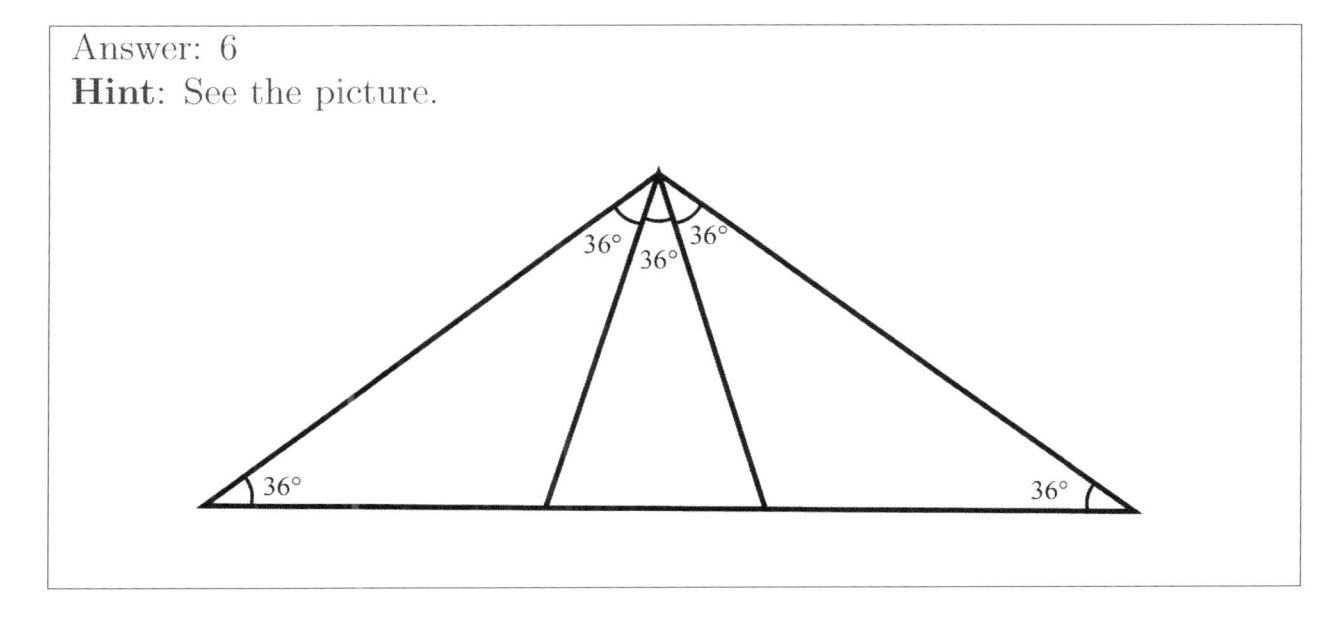

> Answer: 6
> **Hint**: See the picture.

Problem 12: *Let D be a point on hypotenuse AB of right triangle ABC and let K be a point on leg AC. Given that $BD = AC$ and $KC = AD$. Prove that $\angle BAC = 2\angle CBK$.*

> **Hint**: Let E be a point symmetric to K with respect to C, then $AE = AB$.

Chapter 7
Boundary principle

Sometimes, in order to solve a problem it can be helpful to consider some set. It can also be very helpful to consider the largest or the smallest (the rightmost or the leftmost) element (**boundary** element) of that set.

We call this technique **Boundary Principle**.

> **Example 7.1:**
>
> *Andrew* wrote eight numbers in his notebook. We know that there are numbers that are not equal. The sum of any two numbers in that eight numbers is equal to the sum of some other two numbers among the six left. Find the sum of those eight numbers, if we know that the sum of the largest and the smallest is 8.

> **Example 6.1: Solution**
> Answer: 32
> **Solution:** Consider the two greatest of these numbers. According to the condition of the problem, their sum equals to the sum of the remaining two numbers (from these six numbers). Thus, these four numbers are equal.
>
> Similarly, we get that the smallest number is equal to some other three numbers (consider the numbers that their sum is the smallest).
>
> From this fact, we get that the sum of these eight numbers is $4 \cdot 8 = 32$.

7.1 Practice Problems

Problem 7.1. *Prove that the value of $512(1 + \frac{1}{2} + \frac{1}{3} + \cdots + \frac{1}{2022})$ is not an integer.*

Problem 7.2. *Can we split a square with side 20 into 10 squares that have different sides, so that the length of the sides are natural numbers?*

Problem 7.3. *A few football teams took part in a tournament. Any two teams played against each other only once. We say that team A is "stronger" than team B, if either A won against B or there exists a team C so that A won against C and C won against B. Prove that there exists a team in the tournament that is "stronger" than any other team.*

Problem 7.4. *Find a positive integer n not exceeding 19 so that the numbers $1, 2, \ldots n$ can be rearranged (in one line and in any possible order), such that the sum of any two neighbor numbers is a perfect square. For example, if $n = 4$, then one possible rearrangement is 2, 1, 4, 3. Note that $2 + 1 = 3$, $1 + 4 = 5$, $4 + 3 = 7$ and 3, 5, 7 are not perfect squares, so this rearrangement does not work.*

Problem 7.5. *A few volleyball teams took part in a tournament. Any two teams played against each other only once. Prove that there exist four teams A, B, C, D so that team A won against teams B, C, D, team B won against teams C, D, and team C won against team D.*

Problem 7.6. *A few scientists took part in a conference. Some of them were friends with each other. We know that if any two scientists have an equal number of friends (from that conference), then they do not have mutual friends. Prove that there is a scientist that has exactly one friend (from that conference).*

Problem 7.7. *There are 100 nationalities living in an unknown country. A nation X is considered a minority if there can be found 50 nationalities so that the number of their representatives is not less than quadruple of the representatives of nation X. What percent of the population of this unknown country can be classified as a minority?*

Problem 7.8. *Given 100 lines in the plane, no three concurrent (no three pass through the same point) and no two parallel to each other. Prove that these 100 lines divide the plane into such regions that at least 34 of them are triangles.*

Problem 7.9. *Given 30 points on a circle. The segment connecting any two of these points is either blue or red (both colors appear). For any three of these points, if the sides of the triangle that they form have different colors, then we can change the colors of two sides to the color of the other side. Prove that, using the above described procedure, it is possible to recolor all segments so that they all have the same color.*

7.1.1 Solutions of practice problems

Problem 7.1: *Prove that the value of $512(1 + \frac{1}{2} + \frac{1}{3} + \cdots + \frac{1}{2022})$ is not an integer.*

> **Proof**: Note that from $1, 2, \ldots 2022$ only 1024 is divisible by 2^{10}. Thus, the least common multiple of these numbers is $2^{10} \cdot m$, where m is an odd number. It follows that
>
> $$1 + \frac{1}{2} + \cdots + \frac{1}{2022} = \frac{k}{2^{10} \cdot m},$$
>
> where k is odd (k is odd as $2^{10} \cdot m$ is the least common multiple. So, no higher power of 2 can be included there, which means k should be odd). Hence, $512(1 + \frac{1}{2} + \cdots + \frac{1}{2022}) = 512 \cdot \frac{k}{2^{10} \cdot m} = \frac{k}{2 \cdot m}$ is not an integer.

Problem 7.2: *Can we split a square with side 20 into 10 squares that have different sides, so that the length of the sides are natural numbers?*

> Answer: It is not possible.
> **Solution**: Assume the square was divided into 10 such squares. Consider the one whose a side is the largest.
> Consider two cases.
>
> - when $a = 10$
>   In this case, the sides of other squares are $1, 2, \ldots, 9$, thus the sum of the surface areas of these squares is $1^2 + 2^2 + \cdots + 10^2 = 385 \neq 400$, which is impossible.
>
> - when $a \geq 11$
>   In this case, the sum of the surface areas of 10 squares is greater or equal to $1^2 + 2^2 + \cdots + 9^2 + 11^2 = 406$, which is not possible.
>
> Hence our assumption is wrong.

Problem 7.3: *A few football teams took part in a tournament. Any two teams played against each other only once. We say that team A is "stronger" than team B, if either A won against B or there exists a team C so that A won against C and C won against B. Prove that there exists a team in the tournament that is "stronger" than any other team.*

> **Proof**: Assume that team A had the highest number of wins and let us denote that number by n. Prove that team A is "stronger" than all other teams.
> We proceed the proof by contradiction argument.
> Suppose there is a team B so that team A is not "stronger" than team B. This means that at first team B won against team A, then any team that won against A lost against team B. We get that team B has at least $n+1$ wins. This is not possible, as each team cannot win more than n times. This leads to a contradiction.

Problem 7.4: *Find a positive integer n not exceeding 19 so that the numbers $1, 2, \ldots n$ can be rearranged (in one line and in any possible order), such that the sum of any two neighbor numbers is a perfect square. For example, if $n = 4$, then one possible rearrangement is 2, 1, 4, 3. Note that $2+1 = 3$, $1+4 = 5$, $4+3 = 7$ and 3, 5, 7 are not perfect squares, so this rearrangement does not work.*

> **Solution**: Suppose it is possible to rearrange the numbers $1, 2, \ldots, n$ in one line so that the sum of any neighbors is a perfect square. One can easily check that if $n \geq 18$, then the numbers 16 and 18 should be the border numbers of the line. Moreover, 9 should be the neighbor of 16 and 7 should be the neighbor of 18. On the other hand, 16 and 7 should be the neighbors of 9, which is impossible.
> So, n cannot be 18 or 19. For $n = 17$ let us provide an example of numbers that satisfy the assumptions of the problem:
>
> $$16, 9, 7, 2, 14, 11, 5, 4, 12, 13, 3, 6, 10, 15, 1, 8, 17.$$

Problem 7.5: *A few volleyball teams took part in a tournament. Any two teams played against each other only once. Prove that there exist four teams A, B, C, D so that team A won against teams B, C, D, team B won against teams C, D, and team C won against team D.*

> **Proof**: Assume that team A had the highest number of wins. We know that each game ended with a win and the number of games is $\frac{8 \cdot 7}{2} = 28$. Thus, the number of wins is 28. Then, team A has at least 4 wins, as $\frac{28}{8} > 3$.
> Suppose A won against teams B, C, D, E. If we consider only the games played between teams B, C, D, E (the number of these games is 6), then we get that team B (that has the highest number of wins) has at least 2 wins.
> Let team B won against teams C and D. Without loss of generality one can assume that team C won against team D (as one of them won against the other one). This ends the proof.

Problem 7.6: *A few scientists took part in a conference. Some of them were friends with each other. We know that if any two scientists have an equal number of friends (from that conference), then they do not have mutual friends. Prove that there is a scientist that has exactly one friend (from that conference).*

> **Proof**: Consider the scientist A that has the highest number of friends in the conference. Denote that number by n. Note that, according to the condition of the problem any two friends (among n friends of A) have different friends (among the scientists that are at the conference). The number of friends these people have is $1, 2, \ldots, n$, as everyone has at least one friend. That is A and A does not have more than n friends (according to the definition of the number n). This ends the proof.

Problem 7.7: *There are 100 nationalities living in an unknown country. A nation X is considered a minority if there can be found 50 nationalities so that the number of their representatives is not less than quadruple of the representatives of nation X. What percent of the population of this unknown country can be classified as a minority?*

> Answer: 20.
>
> **Solution**: Consider the nationality that has the largest population among the national minorities. Denote that number by n.
>
> The number of people S that belong to that population is not more than $50n$, and the remaining population S_1 is not less than $4n \cdot 50 = 200n$.
>
> The $p = \frac{S}{S+S_1} \cdot 100$ percent of the country's population belongs to national minorities. So
>
> $$p = \frac{S}{S+S_1} \cdot 100 \leq \frac{S}{S+200n} \cdot 100 \leq \frac{50n}{50n+200n} \cdot 100 = 20.$$
>
> Thus, $p \leq 20$.
>
> In the case, when 50 nationalities have one representative and 50 nationalities have four representatives, then $p = 20$.

Problem 7.8: *Given 100 lines in the plane, no three concurrent (no three pass through the same point) and no two parallel to each other. Prove that these 100 lines divide the plane into such regions that at least 34 of them are triangles.*

> **Proof**: Let line a be one of the 100 lines that we considered. Consider the intersection points of all other lines. Let point A be the closest intersection point to line a (see the picture).
>
> [Figure: lines a, b, c, d with triangle having vertices A, B, C]
>
> Let lines b and c intersect at point A. According to the condition of the problem lines b and c also intersect with line a. Note that, among the remaining 97 lines, there does not exist a line d that intersects the triangle with sides a, b, c. Otherwise, there would be an intersection point that is closer to a than A (either where lines d and b intersect, or where lines d and c intersect).
>
> Now we have that for any line in the 100 lines there exists a part of a plane that is a triangle. Moreover, one side of the triangle belongs to that line. If the lines divide the plane into such pieces where the number of triangles is not more than 33, then according to the Pigeonhole's Principle at least four ($\frac{100}{33} > 30$) should include one side of the same triangle, which is not possible. This ends the proof.

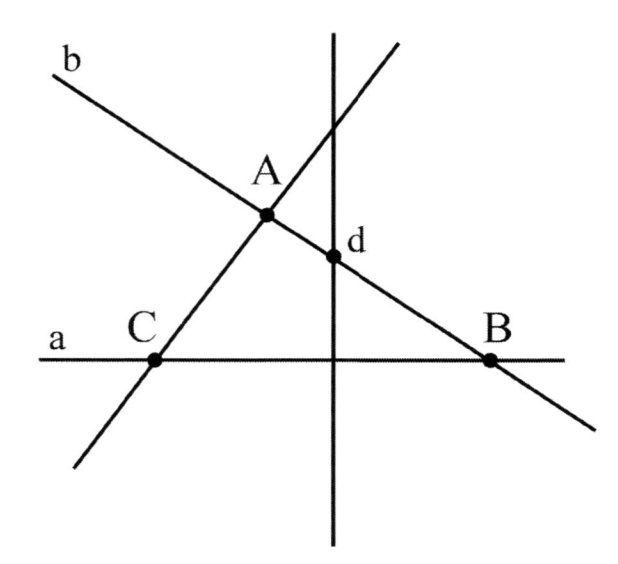

Problem 7.9: *Given 30 points on a circle. The segment connecting any two of these points is either blue or red (both colors appear). For any three of these points, if the sides of the triangle that they form have different colors, then we can change the colors of two sides to the color of the other side. Prove that, using the above described procedure, it is possible to recolor all segments so that they all have the same color.*

Proof: The number of segments is odd $\frac{30 \cdot 29}{2} = 435$. Assume the number of red segments is odd. Prove that it is possible to turn all segments into red. Suppose the number of figures that is possible to get from the given figure is n ($n \leq 2^{435}$). Choose the figure that has the most number of red segments. If that number is not 435 then at least two of the segments are blue. If A and B are endpoints of two different blue segments, then the segment AB is also blue. Indeed, if AB was red then we have another figure with more red segments than the one we took, which is impossible.

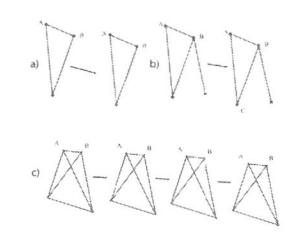

Consider a triangle whose sides are colored with the two colors. It is clear that two sides of the triangle are red, and the other side is blue.

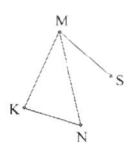

Segments with endpoint M are red, otherwise segment MN is blue. Besides N and K, there is a point S that is an endpoint of a blue segment, as the number of blue segments is even. We have, NS and KS are blue. In that case, we can add the number of red segments, which is impossible.

Chapter 8

Graphs

Consider a finite number of points on a plane. Connect some of those points with segments.

Definition 1. *A figure that is formed from a finite number of points and segments that connect some of these points is called a **graph**.*

Some examples of graphs are given in the figure.

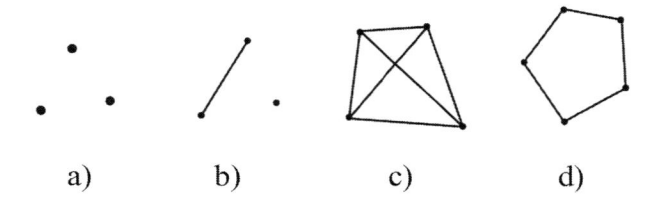

a)   b)   c)   d)

The points are called the **vertices** of a graph, and the segments are called the **edges** of a graph.

Problem 8.1: *Find the number of vertices and edges in the figure above.*

> Answer.
> a) 3 vertices, 0 edges.
> b) 3 vertices, 1 edge.
> c) 4 vertices, 6 edges.
> d) 5 vertices, 5 edges.

Definition 2. *A graph is is called **complete** if there is any edge connecting any two vertices of the graph.*

Only the graph in the figure c) is complete.

Problem 8.2: *Find the number of edges in a complete graph that has:*
a) 2,   b) 3,   c) 6,   d) 10 vertices.

> Answer.
> a) 1 edge
> b) 3 edges
> c) 15 edges
> d) 45 edges

Definition 3. *The degree of a vertex is the number of edges connecting it.*

Problem 8.3: *Find the degrees of all vertices and their sum in the following graph.*

> Answer.
> a) 1, 1, 1, 1, 1, 3
> b) 8

135

Problem 8.4: *Find the sum of the degrees of all vertices of a complete graph that has a) 1, b) 3, c) 15, d) 45   edges.*

> Answer.
> a) 2
> b) 6
> c) 30
> d) 90

Example 8.1:

27 numbers were randomly chosen from the numbers $1, 2, \ldots, 40$. Prove that one of the chosen numbers is the double of another chosen number.

Example 8.1: Proof

Proof: Represent the numbers $1, 2, \ldots, 40$ as points on a plane that are the vertices of the graph. Connect two vertices if one endpoint is double the other. The segments are the edges of the graph.

Consider that graph with some of its edges being red.

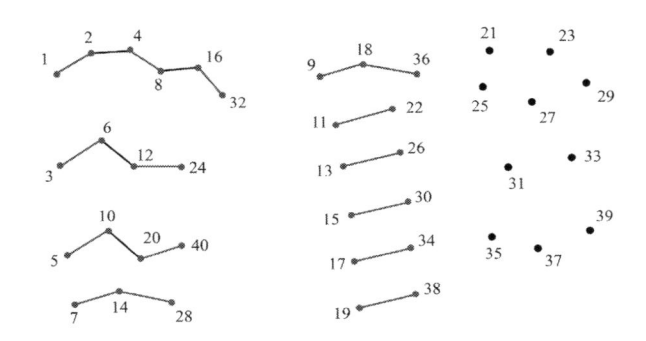

The number of red edges in that graph is 16, and the number of vertices that are not endpoints of a red edge is 10. Thus, according to Pigeonhole's Principle, there can be found two numbers among the 27 chosen so that they are endpoints of a red edge, which means that one is double the other.

There is a relationship between the degrees of vertices and the number of edges in any graph, represented by the following theorem.

Theorem 8.1. *Prove that the for any graph the sum of the degrees of its vertices is twice as much as the number of edges.*

Proof. For each edge of the graph remove some portion of its inner part. The edge now is divided into two "half-edges".

Note that each vertex of the graph is an endpoint for as much "half-edges" as its degree. So the sum of the degrees of all vertices is equal to the number of all "half-edges". The latter is twice as much as the number of edges since each edge was divided into two "half-edges". □

Example 8.2:

The degrees of five vertices of a graph with six vertices are $5, 5, 4, 3, 2$. Find the degree of the sixth vertex.

Example 8.2: Solution

Solution: We know that two vertices of the graph have degree 5, so the degree n of the sixth vertex is not less than 2. From the fact that one of the degrees is 2, we may conclude that $n \leq 4$. According to the theorem $5 + 5 + 4 + 3 + 2 + n = n + 19$ is an even number. Thus considering the fact that $2 \leq n \leq 4$ we get that $n = 3$.
see the picture that shows that $n = 3$ is possible.

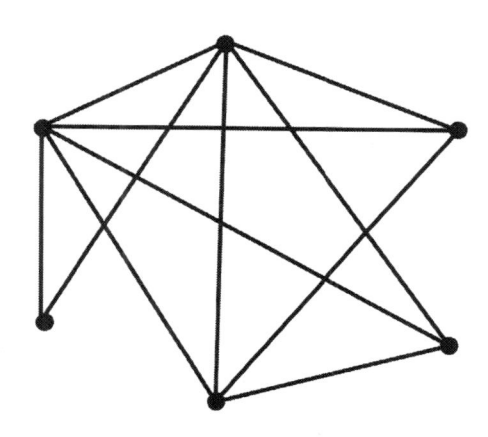

Definition 4. *A graph is called **regular** if all the degrees of its vertices are equal.*

Problem 8.5: *Find the number of edges of a graph with 2022 vertices if the degree of each vertex is a) 1,   b) 2,   c) 3.*

> Answer.
> a) 1011
> b) 2022
> c) 3033

Definition 5. *Two endpoints of a graph that are the endpoints of the same edge are called **adjacent** or **neighboring vertices**.*

It is obvious that the number of neighbors of a vertex is equal to its degree.

> **Example 8.3:**
>
> A mathematics conference has 15 participants. Each participant is friends with at least seven other participants. Prove that any two participants are either friends or have a common friend.

> **Example 8.3: Proof**
> **Proof:** Take 15 points on a plane each representing a participant of a conference. If two participants are friends with each other, then let us connect their points by an edge.
> We get a graph with 15 vertices, where all vertices have degrees of at least seven.
>
> Let A and B be any two vertices of that graph.
>
> If A and B are not neighbors, then it has at least 7 neighbors among the other 13 vertices. Thus, one of those 13 vertices is a neighbor for both A and B.
> This ends the proof.

Definition 6. *A vertex is called **odd** if it has an **odd degree**. Otherwise, it is called **even**.*

From theorem 8.1 we get the following theorem.

Theorem 8.2. *The number of odd degree vertices of any (undirected) graph is always even.*

> **Example 8.4:**
> Consider 15 cities. We say two cities are connected if there is a direct highway between them. Is it possible that each city has a direct connection with exactly five of these cities?

> **Example 8.4: Solution**
> **Solution:** Consider 15 points instead of these 15 cities. Connect any two of these points if there is a direct highway from one of these cities to another. We have a graph where every vertex has an odd degree and the number of vertices is 15. So, according to theorem 8.2 it is not possible.

Definition 7. *A graph is called **connected** if for any two vertices A and B there exist vertices $C, D, E, \ldots L$ so that the following vertices are neighbors A and C, C and D, D and E,…, L and B (see the picture).*

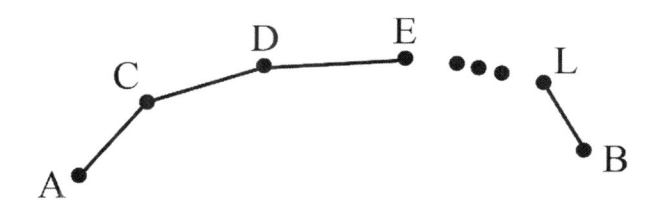

In this case, we say that there exists a path from A to B.
The following figure shows some examples of connected graphs.

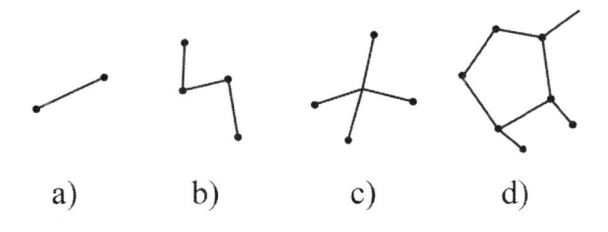

a) b) c) d)

We consider the graph with one vertex as a connected graph. In that case, any graph is made of some subgraphs, each of which is connected. Moreover, any two vertices from different subgraphs are not connected.

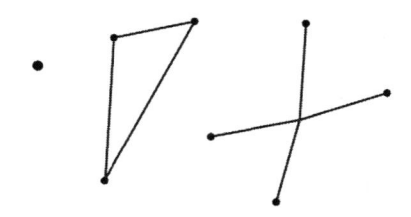

Each of these **sub-graphs** is called a **connected component** of the graph (sub-graph is just any part of the initial graph).

> **Example 8.5:**
>
> Consider nine cities labeled from 1 to 9. Cities a and b are connected by a highway if the two digit number $\overline{ab}$ (or $\overline{ba}$) is divisible by 3. Is it possible to travel from city 1 to city 9 using highways?

> **Example 8.5: Solution**
> **Solution:** Draw the graph representation of the problem. We connect the two cities if there is a highway between them. 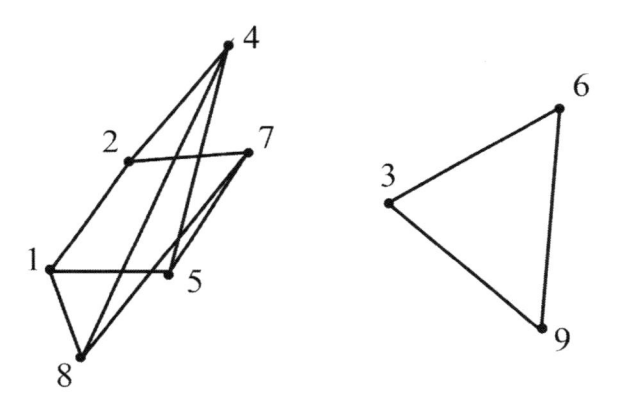
>
> Cities 1 and 9 belong to different connected components. So, it is impossible to go from city 1 to the city 9.

It is possible to prove the following theorem.

Theorem 8.3. *Prove that the number of edges of a graph with n edges is not less than $(n-1)$.*

8.1 Practice Problems

Problem 8.1. *Is it possible to have 15 circles on a plane so that each circle touches exactly 3 other circles?*

Problem 8.2. *Prove that it is possible to find three people among six people so that either three of them know each other or they do not.*

Problem 8.3. *Is it possible to move the knights in first picture (on the left) to obtain the second picture (on the right)?*

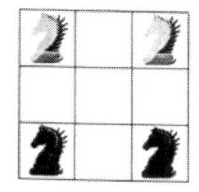

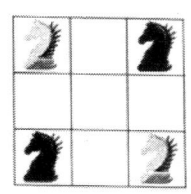

Problem 8.4. *Find the smallest natural number n for which if we choose any n of the numbers from $1, 2, \ldots, 10$ and $21, 22, \ldots, 30$ then one of the numbers is divisible by some other.*

Problem 8.5. *Find the number of nine-digit numbers written using the numbers from $1, 2, \ldots, 9$ so that none has a repeating digit and the number written using any two of its digits is divisible by at least one of the numbers 3, 7 or 13.*

Problem 8.6. *Nine students Ani, Ashot, Gayane, Tigran, Armenuhi, Lusine, Samvel, Hayk, Atom took part in a math competition and only Hayk solved the last problem. He gave the solution only to those students whose name has at least two common letters with his name. After that, each student gave the solution to another student whose name has at least two common letters with the name of that student (if that student did not already know the solution). Did the solution finally get to Ashot?*

Problem 8.7. *The middle square of a 5×5 squared board was removed. Is it possible to pass through all the squares of the remaining board using the steps of the chess knight by visiting the squares only once?*

Problem 8.8. *Find a number n, where $n \geq 4$ so that there exists a graph with n vertices. Moreover, each vertex has degree 3, and any two edges either do not intersect or the intersection point is a vertex of that graph.*

Problem 8.9. *The figure shows a plan of a house without doors and windows. How many doors there should be in the house so that it is possible to go from one room to the other (not necessarily through just one door) (N.M. Sedrakyan)?*

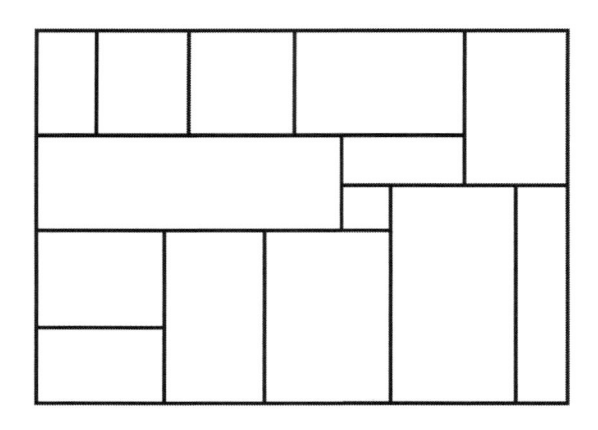

8.1.1 Solutions of practice problems

Problem 8.1: *Is it possible to have 15 circles on a plane so that each circle touches exactly 3 other circles?*

> Answer: It is not possible.
> **Solution**: Instead of each circle take a point on a plane. We have 15 points. We connect two points if their circles touch each other. We get a graph with 15 vertices. According to the condition of the problem, each vertex of the graph has degree 3, which is not possible (by theorem 8.2).

Problem 8.2: *Prove that in any group of six people there are always at least three people who either all know one-another or all are strangers to one-another.*

> **Proof**: Consider six points (instead of six people) on a circle. We connect two points by red edge if these two people know each other. Otherwise, we connect them by a black edge. We need to prove that there is a triangle, so that all its sides are of the same color. Choose a point A. At least three of five edges that have A as an endpoint, are of the same color. For example, we can assume they are red. Name those AB, AC, AD.
>
> a) b)
>
> If any of the edges BC, CD, BD is red, then we have a triangle with red sides. If not, then all sides of triangle BCD are black. Thus, there is a triangle, so that all its sides are of the same color. This ends the proof.

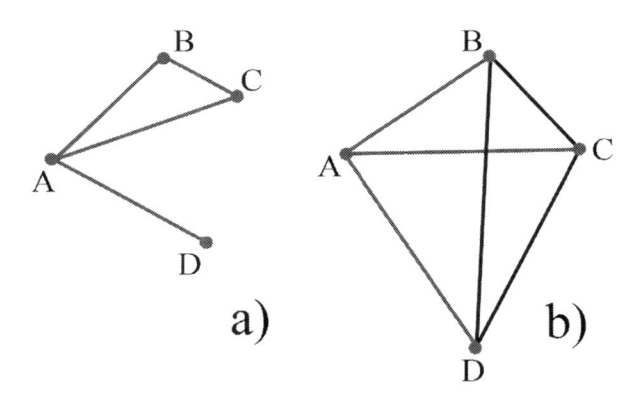

Problem 8.3: *Is it possible to move the knights in first picture (on the left) to obtain the second picture (on the right)?*

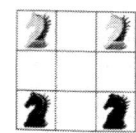

Solution: Write the numbers $1, 2, \ldots, 9$ in a 3×3 board (see the picture).

1	2	3
4	5	6
7	8	9

Instead of each number (square) take a point on a plane. Connect two points if the knight can go from that square to another (we get this graph).

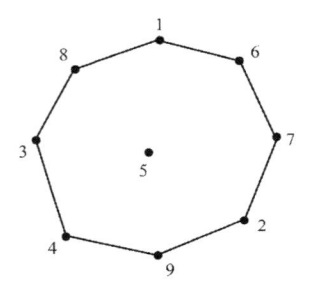

For our problem, we get the following two graphs.

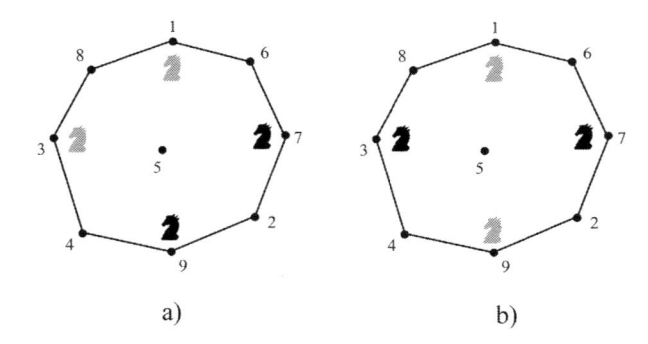

a) b)

Every knight moves from one vertex of the graph to another one (if there is no other knight in that vertex). Clearly, we cannot get figure b) from figure a), as the neighbors of each knight must be of different colors.

Problem 8.4: *Find the smallest natural number n for which if we choose any n of the numbers from* $1, 2, \ldots, 10$ *and* $21, 22, \ldots, 30$ *then one of the numbers is divisible by some other.*

> Answer: 12.
> **Solution**: For each number take a point on a plane, overall 20 points. Connect any two points if its number is divisible by the number of the other point (we get a graph with 20 vertices). Some of the edges of that graph are shown on the figure below.
>
> Note that it is possible to choose 11 numbers, so that none of them is divisible by the remaining 10 numbers. For example
>
> $$6, 8, 9, 10, 21, 22, 23, 25, 26, 28, 29.$$
>
> If we choose 12 numbers, then according to the pigeonhole principle, there are two numbers that belong to the same *connected component* (see the figure). The largest of these numbers is divisible by some other one from this numbers.

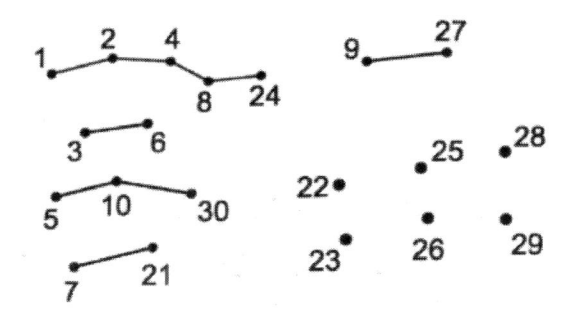

Problem 8.5: *Find the number of nine-digit numbers written using the numbers from $1, 2, \ldots, 9$ so that none has a repeating digit and the number written using any two of its digits is divisible by at least one of the numbers 3, 7 or 13.*

> Answer: 48.
>
> **Solution**: Consider 9 points on a plane. Each number is represented by a point. We will connect two points if the number written using the two numbers as its digits is divisible by either 3, or 7, or 13. This will give the following graph.
>
> Note that in nine-digit number on one side of the number 5 are written the numbers $3, 6, 9$ and on the other side the numbers $1, 2, 4, 7, 8$. Then it follows that the number of such nine-digit numbers is $2 \cdot 2 \cdot 3 \cdot 2 \cdot 5 = 48$.

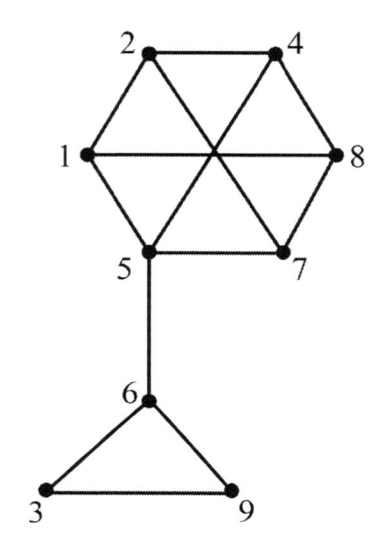

Problem 8.6: *Nine students Ani, Ashot, Gayane, Tigran, Armenuhi, Lusine, Samvel, Hayk, Atom took part in a math competition and only Hayk solved the last problem. He gave the solution only to those students whose name has at least two common letters with his name. After that, each student gave the solution to another student whose name has at least two common letters with the name of that student (if that student did not already know the solution). Did the solution finally get to Ashot?*

> **Proof**: Let these nine points represent these nine students (see the figure). We connect two points if the name of those two student have at least two common letters. We get the graph shown below.
>
> This graph is a connected graph, so Ashot finally got the solution.

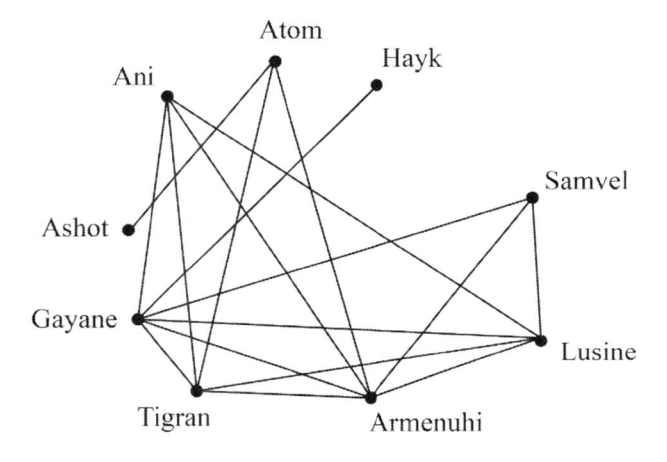

Problem 8.7: *The middle square of a 5×5 squared board was removed. Is it possible to pass through all the squares of the remaining board using the steps of the chess knight by visiting the squares only once?*

Proof: Represent each square of the board by a point on a plane. Assume the knight could pass through all the 24 squares and come back to its initial position. Connect two points if at some point the knight passed from one point to the other. We have a graph with 24 vertices, where the degree of each vertex is two and the graph is connected.

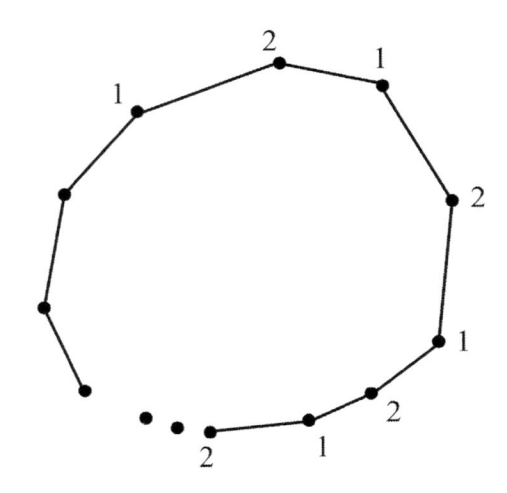

Assume that the knight moves by the edges of the graph, moreover in the clockwise direction. Now consider the squares with numbers 1 and 2.

1				1
		2		
	2		2	
		2		
1				1

Note that from a square with the number 1, the knight can only move to a square with the number 2. Similarly, it can return to a square with the number 1 only from a square with the number 2.

This will lead to the fact that the numbers 1 and 2 alternate each other in the graph representation of the problem, which is impossible.

Problem 8.8: *Find a number n, where $n \geq 4$ so that there exists a graph with n vertices. Moreover, each vertex has degree 3, and any two edges either do not intersect or the intersection point is a vertex of that graph.*

> Answer: When n is an odd number.
> **Proof**: According to theorem 8.2 we have that n is an even number.
>
> An example for an even number n is provided below, where $n \geq 4$.
>
> $n = 4$
>
> $n \geq 6$

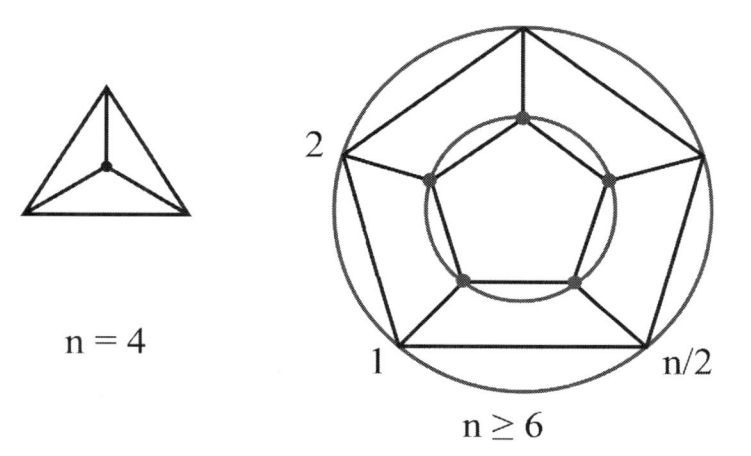

Problem 8.9: *The figure shows a plan of a house without doors and windows. How many doors should there be in the house so that it is possible to go from one room to the other (not necessarily through just one door) (N.M. Sedrakyan)?*

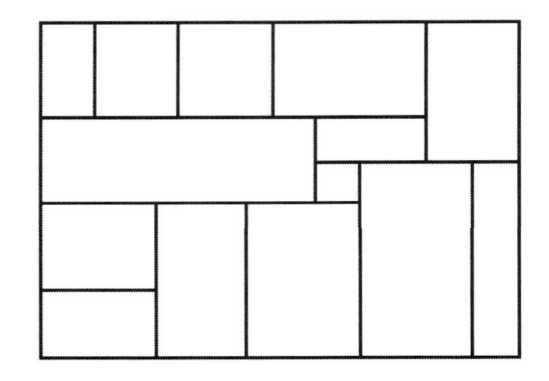

Answer: 14.

Solution: Take a point on a plane for each room in the house, and a point A that belongs to the part outside the house. we get 15 points.

We will connect two of them if the respective rooms are connected with a door. Now we have a graph with 15 vertices connected according to the problem's statement. Note that the number of edges of the graph is equal to the number of doors in the house. According to theorem 3 the number of doors is not smaller than $15 - 1 = 14$.

The figure shows an example of 14 such doors.

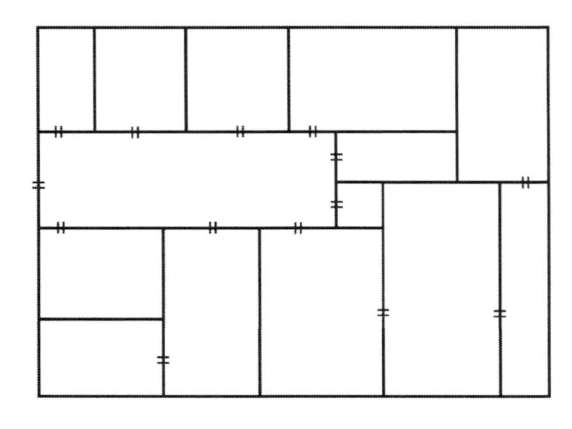

Chapter 9

Solving integer equations

An equation where the value of the *unknown* can only be an **integer** (or a *natural number*) is called an **integer equation**.

In this chapter we return to solving equations, but this time the equations are a bit more advanced. Let us discuss a few examples to understand how the fact that the *unknown* is an *integer* is used to solve *integer equations*.

> **Example 9.1:**
>
> Solve for integers $x^3 + 8x^2 + x - 42 = 0$.

> **Example 9.1: Solution**
> **Answer:** $-7, -2, 2$.
> **Solution:** Let integer x_0 be a solution, thus $42 = x_0(x_0^2 + 8x_0 + 1)$.
> So, x_0 is a divisor of 42. The divisors of 42 are:
> $-42, -21, -14, -7, -6, -3, -2, -1, 1, 2, 3, 6, 7, 14, 21, 42$.
>
> None of $7, 14, 21, 42$ is a solution of this equation, as then we get
>
> $$x_0 > \frac{42}{x_0} = x_0^2 + 8x_0 + 1.$$
>
> None of $-14, -21, -42$ is a solution of this equation, as then we get $\frac{42}{x_0} < 0$ and $x_0^2 + 8x_0 + 1 = x_0(x_0 + 8) + 1 > 1$.
> Plugging in numbers $-7, -6, -3, -2, -1, 1, 2, 3, 6$ into the equation we find out that the solutions are $-7, -3, 2$.

Example 9.2:

Solve the equation with integers: $(x+1)(x+3) = 13$

Example 9.2: Solution
Answer: $(-14, -4), (-2, -16), (0, 10), (12, -2)$.
Solution: We have that if (x_0, y_0) pair is a solution to the problem, where x_0 and y_0 are integers, then $(x_0 + 1)$ is a divisor of 13.

So $x_0 + 1 = -13$ or $x_0 + 1 = -1$ or $x_0 + 1 = 1$ or $x_0 + 1 = 13$. We get

$$\begin{cases} x_0 + 1 = -13 \\ y_0 + 3 = -1 \end{cases} \text{ or } \begin{cases} x_0 + 1 = -1 \\ y_0 + 3 = -13 \end{cases} \tag{9.1}$$

$$\begin{cases} x_0 + 1 = 1 \\ y_0 + 3 = 13 \end{cases} \text{ or } \begin{cases} x_0 + 1 = 13 \\ y_0 + 3 = 1 \end{cases} \tag{9.2}$$

So, the integer solutions for this equation are the pairs $(-14, -4), (-2, -16), (0, 10), (12, -2)$.

Example 9.3:

Solve the equation with integers: $x^2 - y^2 = 2022$

Example 9.3: Solution
Answer: $(506, 504), (106, 96)$.
Solution: We have $x^2 - y^2 = (x - y)(x + y)$ and $(x - y) + (x + y) = 2x$, so $x - y$ and $x + y$ have the same parity. Moreover, $x - y < x + y$, and from the equation it follows that $x - y > 0$.

The canonical form of 2020 is $2^5 \cdot 5 \cdot 101$. So we have:

$$\begin{cases} x - y = 2 \\ x + y = 1010 \end{cases} \text{ or } \begin{cases} x - y = 10 \\ x + y = 202 \end{cases} \tag{9.3}$$

From which we get the pairs $(506, 504), (106, 96)$.

Example 9.4:

Solve the equation with integers: $2xy - 3x - 10y = 3$.

Example 9.4: Solution
Answer: $(3, -3), (-1, 0), (-13, 1), (23, 2), (11, 3), (7, 6)$.
Solution: Write the equation in the following form

$$x(2y - 3) - 10y = 3.$$

Now, let us rewrite this equation in the following form

$$x(2y - 3) - 5(2y - 3) = 3 + 15.$$

Thus, we get that

$$(x - 5)(2y - 3) = 18.$$

The following cases are the possible outcomes:

$$\begin{cases} 2y - 3 = -9 \\ x - 5 = -2 \end{cases} \text{ or } \begin{cases} 2y - 3 = -3 \\ x - 5 = -6 \end{cases} \text{ or } \begin{cases} 2y - 3 = -1 \\ x - 5 = -18 \end{cases} \text{ or } \begin{cases} 2y - 3 = 1 \\ x - 5 = 18 \end{cases} \quad (9.4)$$

$$\begin{cases} 2y - 3 = 3 \\ x - 5 = 6 \end{cases} \text{ or } \begin{cases} 2y - 3 = 9 \\ x - 5 = 2 \end{cases} \quad (9.5)$$

The solutions to the problem are the pairs $(3, -3), (-1, 0), (-13, 1), (23, 2), (11, 3), (7, 6)$.

Some equations have an infinite number of solutions. For others, the solutions are given by formulas.

> **Example 9.5:**
>
> Solve the equation with integers: $x^2 + y^2 = z^2$.

> **Example 9.5: Solution**
> **Solution:** We have $\frac{z-x}{y} = \frac{y}{z+x}$.
> Suppose $\frac{y}{z+x} = \frac{u}{v}$, where u and v are mutually prime.
>
> From the fact $yv = u(z+x)$ we have $y = uk$, in that case, $z + x = vk$, where $k \in \mathbb{N}$.
> From the fact $(z-x)v = yu$ we have $y = vl$, where $l \in \mathbb{N}$ and $z - x = ul$.
> Now we have $uk = y = vl$, so $k = vt, l = ut$, where $t \in \mathbb{N}$.
>
> So $y = uvt, z + x = v^2t, z - x = u^2t$, from which $z = \frac{v^2+u^2}{2}t, x = \frac{v^2-u^2}{2}$, where $v > u$.
>
> When u and v have different parity, in that case, t is even, which means $t = 2d$, where $d \in \mathbb{N}$. we have $x = (v^2 - u^2)d, y = 2uvd, z = (v^2 + u^2)d$, where $u, v, d \in \mathbb{N}$ and $(u, v) = 1$.
> When u and v have the same parity $u + v = 2v_1, v - u = 2u_1$, where $(u_1, v_1) = 1$ and $x = 2u_1v_1t, z = (u_1^2 + v_1^2)t, y = (v_1^2 - u_1^2)t$.
>
> So, the solutions (x, y, z) are $((v_1^2 - u_1^2)d, 2uvd, (u_1^2 + v_1^2)d)$, $(2uvd, (v_1^2 - u_1^2)d, (u_1^2 + v_1^2)d)$, where $u, v, d \in \mathbb{N}$ and $(u, v) = 1$.

9.1 Practice Problems

Problem 9.1. *Find all the pairs of two-digit numbers $\overline{ab}$ and $\overline{cd}$ so that for each of them $\overline{abcd}$ is divisible by $\overline{ab} \cdot \overline{cd}$.*

Problem 9.2. *Find the integer solutions to the following problem.*
$$4x^2 + y^2 + 5 = 12x + 2y.$$

Problem 9.3. *Find the integer solutions to the following problem.*
$$20x + 23y = 2023.$$

Problem 9.4. *Find the integer solutions to the following problem.*
$$20u^2 + 23v^2 = 2023.$$

Problem 9.5. *Is it possible that the difference between two numbers of form $n^2 + 2022n$ is 2026 if n is a natural number?*

Problem 9.6. *Ann wrote some integers on the board. Then Greg wrote the squares of these numbers. After that Sam added all the numbers (squares and the initial numbers) and got 2023. Did Sam count correctly?*

Problem 9.7. *Find two integer solutions of the following equation.*
$$x^2 + (x+1)^2 = y^2.$$

Problem 9.8. *Find the integer solutions of the following equation.*
$$x(x+5) = 4y(y+1).$$

Problem 9.9. *Find the integer solutions of the following equation.*
$$p^2 = q^2 + r^2 + s^2 + 6.$$

9.1.1 Solutions of practice problems

Problem 9.1: *Find all the pairs of two-digit numbers $\overline{ab}$ and $\overline{cd}$ so that for each of them $\overline{abcd}$ is divisible by $\overline{ab} \cdot \overline{cd}$.*

> Answer: $(13, 52), (17, 34)$.
> **Solution**: According to the problem's statement, we have
> $$\overline{abcd} = \overline{av} \cdot \overline{cd} \cdot k$$
> , where $k \in \mathbb{N}$. We have
> $$\overline{ab00} + \overline{cd} = \overline{ab} \cdot \overline{cd} \cdot k.$$
> $$100 \cdot \overline{ab} + \overline{cd} = \overline{ab} \cdot \overline{cd} \cdot k.$$
> $$\overline{cd} = \frac{100\overline{ab}}{k\overline{ab} - 1}.$$
> Note that $\overline{ab}$ and $k\overline{ab} - 1$ are mutually time and from the equation for $\overline{cd}$ we get that $100 \vdots (k\overline{ab} - 1)$.
>
> Thus, we get
> $$\begin{cases} k\overline{ab} - 1 = 20 \\ \overline{cd} = 5\overline{ab} \end{cases} \quad or \quad \begin{cases} k\overline{ab} - 1 = 25 \\ \overline{cd} = 4\overline{ab} \end{cases} \tag{9.6}$$
>
> $$\begin{cases} k\overline{ab} - 1 = 50 \\ \overline{cd} = 2\overline{ab} \end{cases} \quad or \quad \begin{cases} k\overline{ab} - 1 = 100 \\ \overline{cd} = \overline{ab} \end{cases} \tag{9.7}$$
>
> From here we get.
> $$\begin{cases} \overline{ab} = 13 \\ \overline{cd} = 52 \end{cases} \quad or \quad \begin{cases} \overline{ab} = 17 \\ \overline{cd} = 34 \end{cases} \tag{9.8}$$

Problem 9.2: *Find the integer solutions to the following problem.*

$$4x^2 + y^2 + 5 = 12x + 2y.$$

Answer: $(2,3), (1,3), (2,-1), (1,-1)$.
Solution: We have

$$4x^2 - 12x + 9 + y^2 - 2y + 1 = 5.$$

$$(2x-3)^2 + (y-1)^2 = 1^2 + 2^2.$$

Thus,

$$\begin{cases} 2x - 3 = 1 \\ y - 1 = 2 \end{cases} \text{ or } \begin{cases} 3 - 2x = 1 \\ y - 1 = 2 \end{cases} \quad (9.9)$$

$$\begin{cases} 2x - 3 = 1 \\ 1 - y = 2 \end{cases} \text{ or } \begin{cases} 3 - 2x = 1 \\ 1 - y = 2 \end{cases} \quad (9.10)$$

The integer solutions for these equations are $(2,3), (1,3), (2,-1), (1,-1)$.

Problem 9.3: *Find the integer solutions to the following problem.*

$$20x + 23y = 2023.$$

Answer: The pairs $23l + 100, 1 - 20l$, where $l \in \mathbb{Z}$.
Solution: Write the equation in the following form $20(x-100) = 23(1-y)$, so $x - 100 \vdots 23$. From here we have that $x - 100 = 23l, 1 - y = 20l$, where l is any integer. Thus the solution for this equation are the pairs $23l + 100, 1 - 20l$, where $l \in \mathbb{Z}$.

Problem 9.4: *Find the integer solutions to the following problem.*
$$20u^2 + 23v^2 = 2023.$$

> Answer: $(10, 1), (-10, 1), (10, -1), (-10, -1)$.
> **Solution**: Denote $u^2 = x$ and $v^2 = y$ and we get Problem 3, moreover $x \geq 0, y \geq 0$.
>
> Note that we get the integer solutions of Problem 3, when $l \in \{-4, -3, -2, -1, 0\}$. Thus, the pairs (u, v) are $(10, 1), (-10, 1), (10, -1), (-10, -1)$.

Problem 9.5: *Is it possible that the difference between two numbers of form $n^2 + 2022n$ is 2026 if n is a natural number?*

> Answer: It is not possible.
> **Solution**: According to the problem's statement we need to figure out if there are natural numbers n and m so that
> $$n^2 + 2022n - (m^2 + 2022m) = 2026.$$
> From which we get that $(n - m)(n + m + 2022) = 2026$. This is not possible since $(n - m)$ and $(n + m + 2022)$ have the same parity. So their product is either an odd number, or is divisible by 4, but 2026 is not such.

Problem 9.6: *Ann wrote some integers on the board. Then Greg wrote the squares of these numbers. After that Sam added all the numbers (squares and the initial numbers) and got 2023. Did Sam count correctly?*

> Answer: No.
> **Solution**: Note that if a is an integer $a^2 + a = a(a + 1)$ is a product of two consecutive numbers, which is even. Thus the sum of any integer and its square that is written on the board should be even. From this follows that the sum of all numbers is also even. This means that Sam should have gotten an even number.

Problem 9.7: *Find two integer solutions of the following equation.*
$$x^2 + (x+1)^2 = y^2.$$

> Answer: $(3, 5)$, $(20, 29)$.
> **Solution**: One can look for the possible solutions in the list of the *Pythagorean triples*. Checking out this list, it is obvious that the following two solutions $(3, 5)$ and $(20, 29)$ satisfy.
> **Alternative solution**. According to Example 9.5, if we choose $x + 1 = m^2 - n^2$ and $x = 2mn$, where m, n are natural numbers, then $y = m^2 + n^2$. We need to find the pairs (m, n) for which the equation $m^2 - n^2 - 2mn = 1$ is correct. This equation can be rewritten as:
> $$m^2 = \frac{(m+n)^2 + 1}{2}.$$
> For $m + n = 7$ and $m + n = 41$ we get $m = 5$ and $m = 29$, respectively. So, an example of such (x, y) pairs are $(20, 29), (696, 985)$.

Problem 9.8: *Find the integer solutions of the following equation.*
$$x(x+5) = 4y(y+1).$$

> Answer: $(3, 2)$.
> **Solution**: Write the given equation in the following form.
> $$4x^2 + 20x + 25 = 16y^2 + 16y + 4 + 21,$$
> $$(2x+5)^2 - (4y+2)^2 = 21,$$
> $$(2x + 4y + 7)(2x - 4y + 3) = 21.$$
> Note that $2x + 4y + 7 \geq 2 + 4 + 7 = 13$, thus
> $$2x - 4y + 3 = 1,$$
> $$2x + 4y + 7 = 21.$$
> From here we get that $8y + 4 = 20$, thus $y = 2, x = 3$.

Problem 9.9: *Find the integer solutions of the following equation.*

$$p^2 = q^2 + r^2 + S^2 + 6.$$

Answer: $(7,5,3,3), (7,3,5,3), (7,3,3,5)$.
Solution: If none of the numbers q, r, S is divisible by 3 then dividing each of the numbers q^2, r^2, S^2 by 3 we get a remainder 1. Thus, $q^2 + r^2 + S^2 + 6$ is divisible by 3. Form this we get that p is divisible by 3, thus $p = 3$, which is impossible. We get that one of the numbers q, r, S is divisible by 3, thus it is equal to 3.

Assume $q = 3$.
If none of the numbers r and S is divisible by 3, then dividing $r^2 + S^2 + 150$ by 3 we get a remainder 2. Thus it cannot be a square of a natural number.
So, $r = 3$ or $S = 3$.
Assume $r = 3$. we have

$$(p - S)(p + S) = 24.$$

From here

$$\begin{cases} p - S = 2 \\ p + S = 12 \end{cases} \quad (9.11)$$

This is because $(p + S) - (p - S) = 2S \geq 4$. So $p = 7, S = 5$.

Chapter 10
Invariants

Consider an object A and a **transformation**. If we apply that *transformation* on A we get object B, then applying the same transformation on B we get object C, and so on. Usually we see problems of the following form: given two objects A, F and a *transformation*, where we need to figure out is it possible to get F by applying given *transformation* on A.

To solve this kind of problems we either mention a sequence $A, B, C, \ldots, F$ that is a result of applying given *transformation* on the previous element of the sequence to get the next one, or we describe an entity which is fixed (is **invariant**) for all objects $A, B, C, \ldots, F$.

If an *invariant* is different for two objects A and F it is clear that we cannot get object F from object A by using the given transformation.

It is more difficult to find the invariants of the given *transformation*, so that from the equality of objects A and F it follows that it is possible to transform A to F.

These *invariants* are called **full system of invariants**.

We can consider several *invariants* instead of one (see this example).

Example 10.1:

Numbers $1, 2, 3, 4, 5, 6$ are written on the board. *Andrew* chooses two of these numbers and decreases one by 1 and increases the other by 3. After that he erases the previous two numbers. Is it possible that after this the final numbers on the board are $1, 2, 3, 2021, 2022, 2023$?

Example 10.1: Solution
Answer: It is not possible.
Solution: Note that after each transformation the sum of the numbers on the board is increasing by 2. So the after each step the parity of the sum of the numbers is not changed. The sum of the numbers $1, 2, 3, 4, 5, 6$ is odd, but the sum of the numbers $1, 2, 3, 2021, 2022, 2023$ is even.

Example 10.2:

Ten coins are placed on a circle. It is possible to turn four consecutive coins or choose one coin, turn two consecutive coins to the positive direction and the other two coins to the negative direction. Is it possible to turn all the coins using this procedure?

Example 10.2: Solution
Answer: It is not possible.
Solution: Instead of the coins let us write ten ones. If we turn a coin, then we replace its number with its opposite.

At the beginning, we have ten ones and after some steps we need to have ten negative ones. Note that at each step two of the chosen four numbers are the numbers at the white dots. We replace these two numbers along with two others with their opposites. Thus, we get that the product of the numbers at two white dots is invariant (stays the same at any step). For the first ten numbers it is one. For the second set of numbers it is $-1 \cdot (-1) \cdot (-1) \cdot (-1) \cdot (-1) = -1$. So, it is not possible to get the second ten numbers from the first ten numbers.

10.1 Practice Problems

Problem 10.1. *The numbers $1, 2, 3, 4, 5, 6$ are written on the board. At any step we choose two numbers a and b from the board, erase them and replace them by the number $|a-b|$. After five steps there is only one number left on the board. Is it possible that the number left is:*
a)   2?
b)   5?

Problem 10.2. *There are six piles of nuts on a table. In each pile there are $1, 2, 3, 4, 5, 6$ nuts, respectively. At any step we choose any:*
a)   4
b)   5
piles and add to each of the chosen piles a nut. Is it possible that after some steps the number of the nuts in every pile is the same?

Problem 10.3. *There are seven piles of nuts on a table. At any step we choose 5 random piles and add one nut to each of them. Prove that after a few transformations the number of nuts in all piles can be the same.*

Problem 10.4. *In every square of a 3×3 board is written 0. We transform the board by the following transformation: we either add or subtract 1 to all the numbers in one of its columns (or in one of its rows). Describe all possible boards that we can get after a few such transformations.*

Problem 10.5. *Give a system of invariants for the transformation in problem 13.4.*

10.1.1 Solutions of Practice Problems

Problem 10.1: *The numbers $1, 2, 3, 4, 5, 6$ are written on the board. At any step we choose two numbers a and b from the board, erase them and replace them by the number $|a-b|$. After five steps there is only one number left on the board. Is it possible that the number left is:*
a)   2?
b)   5?

> Answer: a) it is not possible.
> Answer: b) it is possible.
> **Solution**: Note that $a + b$ and $|a - b|$ have the same parity.
>
> So after each step the parity of the sum of all numbers written on the board will not change.
>
> In the beginning, the sum of the numbers is $1 + 2 + 3 + 4 + 5 + 6 = 21$, which is odd. It follows that after five transformations the sum should be odd, so
>
> a) 2 is not possible
>
> b) 5 is possible.
>
> For example, if we use these steps:
>
> $$1, 2, 3, 4, 5, 6 \to 2, 3, 4, 5, 5 \to 1, 4, 5, 5 \to 1, 1, 5 \to 0, 5 \to 5.$$

Problem 10.2: *There are six piles of nuts on a table. In each pile there are $1, 2, 3, 4, 5, 6$ nuts, respectively. At any step we choose any:*
a) 4
b) 5
piles and add to each of the chosen piles a nut. Is it possible that after some steps the number of the nuts in every pile is the same?

> Answer: a) it is not possible.
> Answer: b) it is possible.
> **Solution**: Suppose after some steps the number of nuts in the piles is a, b, c, d, e, f. Consider the case when $a \leq b \leq c \leq d \leq e \leq f$.
>
> When $a < f$, separate two cases.
>
> - $e < f$.
>   After the step $a + 1, b + 1, c + 1, d + 1, e + 1, f$ we have that $a + 1 \leq b + 1 \leq c + 1 \leq d + 1 \leq e + 1 \leq f$ and $f - (a + 1) < f - a$.
>
>   This means that after this step, the difference between the largest and the smallest numbers became less.
>
> - $e = f$.
>   After the step $a+1, b+1, c+1, d+1, e+1, f$ we have that the difference between the largest and the smallest numbers is not changed, but the number of the numbers equal to the largest number decreased by one.
>
> It is clear that after some such steps, we have the smallest and the largest numbers equal.
>
> That is we have the same number of nuts in all the piles.

Problem 10.3: *There are seven piles of nuts on a table. At any step we choose 5 random piles and add one nut to each of them. Prove that after a few transformations the number of nuts in all piles can be the same.*

> **Solution**: Assume the number of the nuts in the piles is initially $a \leq b \leq c \leq d \leq e \leq f \leq g$. we get the following steps.
>
> $$a, b, c, d, e, f, g \to$$
> $$a+1, b+1, c+1, d+1, e+1, f, g \to$$
> $$a+2, b+2, c+2, d+1, e+1, f+1, g+1 \to$$
> $$a+3, b+2, c+2, d+2, e+2, f+2, g+2 \to$$
> $$a+3, b+3, c+3, d+3, e+3, f+3, g+2$$
>
> .
>
> Thus we exchanged the numbers a, b, c, d, e, f, g numbers with $a+3, b+3, c+3, d+3, e+3, f+3, g+2$.
>
> Thus the difference between the largest and smallest numbers will decrease, or if that number remains the same, then the number of the numbers equal to the largest number will decrease by one. It is clear that after such steps the difference between the largest and the smallest number is 0. Thus, we have an equal number of nuts in all the piles.

Problem 10.4: *In every square of a 3×3 board is written 0. We transform the board by the following transformation: we either add or subtract 1 to all the numbers in one of its columns (or in one of its rows). Describe all possible boards that we can get after a few such transformations.*

> **Solution**: Note that after some such transformations from the table filled only with 0's we get the following table, where a, b, c, d, e are integers.
>
> | a | b | c |
> |---|---|---|
> | d | | |
> | e | | |
>
> At first, we transform by the three columns and we get a, b, c. After this, we get d, e by the two rows. Now let us prove that we can choose all the numbers in the remaining squares with only one way. It is sufficient to note that, for example, in the following figure for the numbers x, y, z, t the expression $x - y + z - t$ is invariant with respect to the given transformation.
>
> | x | y | |
> |---|---|---|
> | t | z | |
> | | | |
>
> Thus, it follows that the remaining four numbers are $b+d-a, c+d-a, b+e-a$ and $c+e-a$.
>
> | a | b | c |
> |---|-------|-------|
> | d | b+d−a | c+d−a |
> | e | b+e−a | e+c−a |
>
> Thus, the tables derived from a table filled with 0's have this form, where a, b, c, d are integers.

Problem 10.5: *Give a system of invariants for the transformation in problem 13.4.*

Solution: According to the solution of the problem 13.4, the expression $x - y + z - t$ is invariant with respect to the given transformation (see the figure).

x	y	
z	t	

x		y
z		t

x	y	
z	t	

x		y
z		t

We mentioned four invariants. From the solution of the problem 13.4, it follows that from this table

a_1	b_1	c_1
d_1		
e_1		

it is possible to get the following table

a	b	c
d		
e		

After filling in the first table (considering four invariants) there is only one way of filling in the second table. This means that four invariants are a *full system of invariants*.

Chapter 11

Mathematics in games

Let us consider a **two-player game**, so that players take turns to play (play one after another). In this chapter we study *games*, where it is possible to find out (describe) the **outcome** (final result) of the game.

So, we need to find out the *winner* and describe the **strategy** (plan of moves) that the *winner* can use to win (so-called **optimal strategy**).

In some games, the *outcome* (final result) for one player can be the same no matter what is the *strategy* (moves) of the other player, for example:

> **Example 11.1:**
> Two people take turns while playing a game at a table. At this table, there is a pile of 100 nuts. The first player splits this into two non-zero piles. From here on out, the players must choose any of the piles on a table and split it into two more non-zero piles. If a player can no longer make a turn, that person loses. Who will win if both players play with optimal strategies?

> **Example 10.1: Solution**
> Answer: The first player.
> **Solution:** Note that no matter what is the strategy, after each move, the number of piles increases by one. So, in the end we have 100 piles (each having one nut). Then, the number of moves is 99. This means that the last move belongs to the first player. So, the first player will finally win.

For this kind of games, we need **to prove** that the winner is the first or the second player, **independent** from the moves (*strategy*) of the other player.

To prove this we need to suggest a **rule** (strategy). Such *rule* can be created, for example, using the concept (idea) of **symmetry**.

> **Example 11.2:**
>
> Two people take turns placing dimes (ten-cent coin) on a table, so that no coin overlaps another. If a player can no longer place a dime on a table, that person loses. Who will win if both players play with optimal strategies?

> **Example 10.1: Solution**
> Answer: The the first player.
> **Solution:** The first player will win. At first, the first player can place the first coin in the center of the table.
>
> Then, after the second player plays, the first player can place a coin symmetric (with respect to the center of the table) to that coin.
>
> This means that anytime the second player plays the first player can always still play.
>
> So, finally the first player will win.

It is also possible to create such a **rule** (strategy) using the concept (idea) of **dividing into pairs**.

> **Example 11.3:**
>
> A figure is placed on the leftmost square of a 1×40 board (split into 40 squares). Let numbers 0 to 39 be written on a board. Two players take turns moving this figure either left or right by any number on the board. Once they move the figure, they erase that number and they cannot use it anymore. If a player moves the figure out of the board, then that player loses. Who will win if both players play with optimal strategies?

> **Example 10.3: Solution**
> Answer: The first player.
> **Solution:** At first, the first player can move the figure by 0 squares.
>
> Let us divide the numbers $0, 1, 2, \ldots, 39$ into the following pairs $(0, 39), (1, 38), (2, 37), \ldots, (19, 20)$.
>
> If the second player moves the figure by n squares, then the first player can move the figure by $39 - n$ squares (in the same direction as the second player).
>
> So, after second player's turn the first player can always place the figure on the rightmost or on the leftmost square of this 1×40 chessboard.
>
> So, the first player will finally win (make the last move).

It is also possible to create such a **rule** (strategy) by creating some **pattern**.

Let us discuss this case in the next example.

Example 11.4:

A figure is placed on the bottom left square of a 8×8 board. Two players take turns moving the figure either right or up one square. If a player moves the figure outside the board, then that player loses. Who will win if both players play with optimal strategies?

Example 10.4: Solution

Answer: The second player.

Solution: Note that, after each move of the first player, the second player can move the figure to one of the red squares (see the picture).

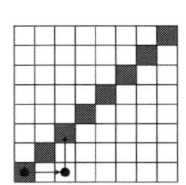

In some cases, the **winning strategy** can be described by moving backward *from the end of the game to the beginning*. For example:

Example 11.5:

Two players take turns writing numbers on the board. The first player writes 1. The next player either writes $a + 1$ or $2a$, if that number is not on the board and if a is written on the board. Whoever writes 18 wins and if a player writes a number greater than 18, then that player loses. Who will win if both players play with optimal strategies?

Example 10.5: Solution

Answer: The first player.

Solution: Label the numbers by W (winning number) or L (losing number). So, 18 is W. This list shows that the winner is the first player, as the first player starts with 1 and then the first player can play 4.

1	2	3	4	5	6	7	8	9	10	11	12	13	14	15	16	17	18
L	L	W	L	L	L	L	W	L	W	L	W	L	W	L	W	L	W

11.1 Practice Problems

Problem 11.1. *Two players take turns placing rooks on a chessboard, so that no player's piece can be attacked by any of the other player's pieces. Only one rook can be on each square. A player loses if cannot make a move anymore. Who will win if both players play with optimal strategies?*

Problem 11.2. *A figure is placed on the bottom left square of a 5×7 board. Two players take turns moving the figure either right one square or up one square. If a player makes a move outside the board, then that player loses. Who will win if both players play with optimal strategies?*

Problem 11.3. *Andrew and Jane take turns placing figures on a chessboard. Only one figure can be on each square. Andrew places the first figure. After Andrew's each turn Jane places her figure in the column where Andrew placed his figure. At his second turn Andrew places his figure in the row where Jane placed her figure at her last turn. Who will win if both players play with optimal strategies?*

Problem 11.4. *Two players take turns placing knights on a 4×4 chessboard, so that no player's piece can be attacked by any of the other player's pieces. Only one knight can be on each square. A player loses if cannot make a move anymore. Who will win if both players play with optimal strategies?*

Problem 11.5. *A figure is placed on the bottom left square of a 8×8 board. Two players take turns moving this figure either right one, two or three squares or move it up one or two squares. Who will win if both players play with optimal strategies?*

Problem 11.6. *Two players take turns placing knights on a chessboard, so that no player's piece can be attacked by any of the other player's pieces. Only one knight can be on a square. The player who cannot make any more moves will lose. Who will win if both players play with optimal strategies?*

Problem 11.7. *Two players take turns writing numbers from a choice of 1, 2, ..., 99 on a board, the same number cannot be written twice. When the sum of all of the written numbers is greater than 550, the player who wrote the last number loses. Who will win if both players play with optimal strategies?*

Problem 11.8. *Two players take turns placing bishops on a chessboard, so that no player's piece can be attacked by any of the other player's pieces. Only one bishop can be on each square. A player loses if cannot make a move anymore. Who will win if both players play with optimal strategies?*

Problem 11.9. *The number 60 is written on a board. Two people play a game where in each turn, a player decreases 60 by one of its divisors and clears the board, writing the new result on the board. The player that writes 0 on the board loses the game. Who will win if both players play with optimal strategies?*

11.1.1 Solutions of Practice Problems

Problem 11.1: *Two players take turns placing rooks on a chessboard, so that no player's piece can be attacked by any of the other player's pieces. Only one rook can be on each square. A player loses if cannot make a move anymore. Who will win if both players play with optimal strategies?*

> Answer: The second player.
> **Solution**: The number of moves is wight. This means that the last step belongs to the second player.

Problem 11.2: *A figure is placed on the bottom left square of a 5 × 7 board. Two players take turns moving the figure either right one square or up one square. If a player makes a move outside the board, then that player loses. Who will win if both players play with optimal strategies?*

> Answer: The first player.
> **Solution**: Note that, at each step, the first player can place his figure in a red square (see the figure).
>
> So, the last move belongs to the first player.

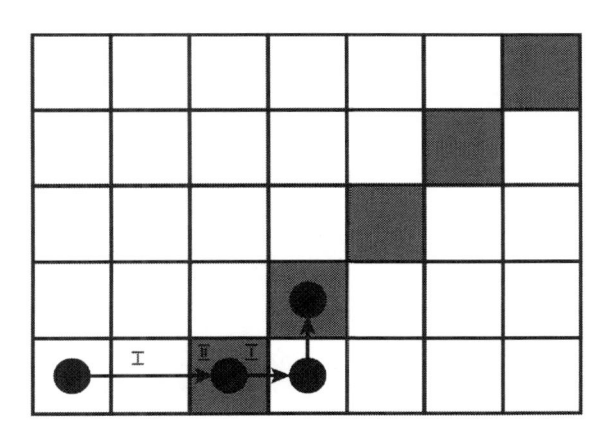

Problem 11.3: *Andrew and Jane take turns placing figures on a chessboard. Only one figure can be on each square. Andrew places the first figure. After Andrew's each turn Jane places her figure in the column where Andrew placed his figure. At his second turn Andrew places his figure in the row where Jane placed her figure at her last turn. Who will win if both players play with optimal strategies?*

> Answer: Jane.
> **Solution**: Note that, Jane can use her "even" turns so that after each turn on each column and on each row there is an even number of figures. So, the last move belongs to Jane.

Problem 11.4: *Two players take turns placing knights on a 4×4 chessboard, so that no player's piece can be attacked by any of the other player's pieces. Only one knight can be on each square. A player loses if cannot make a move anymore. Who will win if both players play with optimal strategies?*

> Answer: The second player.
> **Solution**: Let us divide the squares of this 4×4 chessboard into pairs and number the squares of the same pair by the same number (see the figure).
>
> | 4 | 7 | 6 | 2 |
> |---|---|---|---|
> | 5 | 1 | 3 | 8 |
> | 8 | 3 | 1 | 5 |
> | 2 | 6 | 7 | 4 |
>
> After each step of the first player, the second player can place his knight in the square whose pair was occupied by the knight of the first player in the previous step. So, if the first player can make a move, then the second player can also make a move, as the knights which are on the squares with the same numbers (for example 4 and 4) hit two other knights which are on the squares with the same numbers (for example 3 and 3).

Problem 11.5: *A figure is placed on the bottom left square of a 8 × 8 board. Two players take turns moving this figure either right one, two or three squares or move it up one or two squares. Who will win if both players play with optimal strategies?*

> Answer: The first player.
>
> **Solution**: Let us place + or − sign in each square of the chessboard except the top right square. Moreover, let us place them starting from the top right square, then we will place them in the squares which have a common side with that square.
>
> Sign + means that if the figure is in that square, then the winner is the player whose turn it is. Sign − means the opposite (the other player is the winner).
>
> | + | + | + | − | + | + | + | − |
> |---|---|---|---|---|---|---|---|
> | + | + | − | + | + | + | − | + |
> | + | − | + | + | + | − | + | + |
> | + | + | + | − | + | + | + | − |
> | + | + | − | + | + | + | − | + |
> | + | − | + | + | + | − | + | + |
> | + | + | + | − | + | + | + | − |
> | + | + | − | + | + | + | − | + |
>
> Note that, at each turn the first player can place the figure on a square with − sign.
>
> By doing so, the first player will always win.

Problem 11.6: *Two players take turns placing knights on a chessboard, so that no player's piece can be attacked by any of the other player's pieces. Only one knight can be on a square. The player who cannot make any more moves will lose. Who will win if both players play with optimal strategies?*

Answer: The second player.

Solution: Consider the pairs of squares of the chessboard which are symmetric to each other with respect to the center of the board (see the figure).

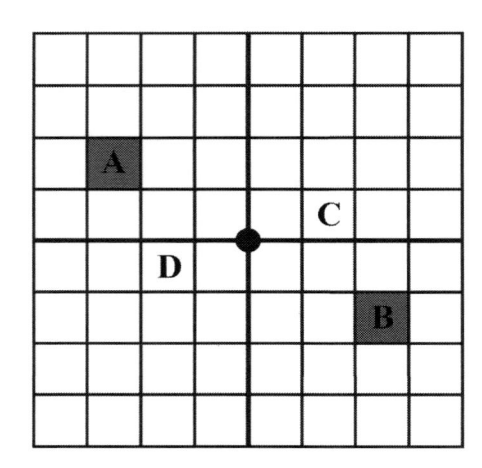

Note that, if the second player places his knight on the square symmetric to the square where the first player placed his knight, then the second player will finally win.

The fact that the second player will finally win can be explained in detail like this: if the first player placed his knight on square A, then the second player can place his knight on square B (see the figure). We need also to check that there is no other knight which hits the knight of the second player (placed on square B).

For example, if there is a knight on square C (which hits the knight of the second player placed on square B), then there is a knight on square D too. This means that the first player would not be able to place a knight on square A. So, there could not be a knight on square C.

So, the last move belongs to the second player.

Problem 11.7: *Two players take turns writing numbers from a choice of 1, 2, ..., 99 on a board, the same number cannot be written twice. When the sum of all of the written numbers is greater than 550, the player who wrote the last number loses. Who will win if both players play with optimal strategies?*

> Answer: The first player.
>
> **Solution**: At first, let us remove 50 from the list $1, 2, \ldots 99$.
>
> Now, let us divide the remaining numbers into 49 pairs of numbers, so that sum of each pair of numbers is 100.
>
> In other words, we consider the pairs $(k, 100 - k)$, where $k = 1, 2, \ldots 49$.
>
> Assume that, at the first move the first player chooses 50.
>
> Then, if the second player chooses number k, the first player will choose number $100 - k$.
>
> If, for example the second player at the first few turns chose the numbers a, b, c, d, e, then the first player chose the numbers $100 - a$, $100 - b, 100 - c, 100 - d, 100 - e$. Note that
>
> $$50 + a + 100 - a + b + 100 - b + c + 100 - c + d + 100 - d + e + 100 - e = 550.$$
>
> So, after a few turns the first player can make the sum of the numbers to become equal to 550.
>
> Thus, the second player makes the sum to become greater than 550.
>
> So, the first player will win.

Problem 11.8: *Two players take turns placing bishops on a chessboard, so that no player's piece can be attacked by any of the other player's pieces. Only one bishop can be on each square. A player loses if cannot make a move anymore. Who will win if both players play with optimal strategies?*

> Answer: The second player.
>
> **Solution**: Consider the pairs of squares of the chessboard which are symmetric to each other with respect to the center of the board (see the figure).
>
> Note that, if the second player places his bishop on the square symmetric to the square where the first player placed his bishop, then the second player will finally win.
>
> The fact that the second player will finally win can be explained like this: if the first player placed his bishop on square A, then the second player can place his bishop on square B (see the figure). We need also to check that there is no other bishop which attacks the bishop of the second player (placed on square B).
>
> For example, if there is a bishop on square C (which hits the bishop of the second player placed on square B), then there is a bishop on square D too. This means that the first player would not be able to place a bishop on square A. So, there could not be a bishop on square C.
>
> So, the last move belongs to the second player.

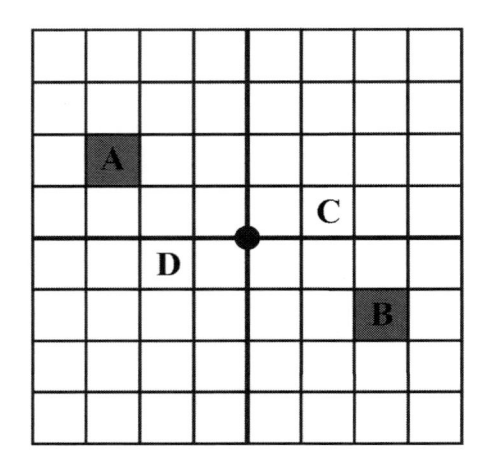

Problem 11.9: *The number 60 is written on a board. Two people play a game where in each turn, a player decreases 60 by one of its divisors and clears the board, writing the new result on the board. The player that writes 0 on the board loses the game. Who will win if both players play with optimal strategies?*

> Answer: The first player.
>
> **Solution**: Let us write the numbers either with a line under a number or write them in bold. We use a line under a number for the numbers that are possible wins and we write the losing numbers in bold. Then, we have the following:
> $$\mathbf{0}, \underline{1}, \mathbf{2}, \underline{3}, \mathbf{4}, \underline{5}, \mathbf{6}, \ldots, \mathbf{44}, \underline{45}.$$
> Thus, it follows that the winner is the first player.
>
> It is possible to explain the solution in the following way.
>
> At the first move, the first player gets the number $60 - 15 = 45$. After this, the second player will get an even number. If the number is not 0, then the first player will turn it to an odd number, and so on. It is clear that, at some step, the second player will get 0.

Chapter 12

Proving inequalities

(a) **The square of any (real) number is non-negative.**
At first, we consider inequalities where we use the fact that the square of any (real) number is non-negative.

In this chapter, all considered numbers (a, b,...) are real numbers.

> **Example 12.1:**
> Prove that
> $$a^2 + b^2 \geq 2ab.$$

Example 12.1: Proof
Proof: Note that $a^2 + b^2 - 2ab = (a-b)^2$. As the square of any (real) number is non-negative, then we have
$$(a-b)^2 \geq 0.$$

So $a^2 + b^2 - 2ab = (a-b)^2 \geq 0$. Thus $a^2 + b^2 \geq 2ab$.

The equality holds when $(a-b)^2 = 0$, that is when $a = b$.

So $a^2 + b^2 = 2ab$ if and only if $a = b$.

Example 12.2:

Prove that
$$a^2 + b^2 + \frac{1}{b^2} + 2 \geq 2 \cdot \frac{a}{b} + 2ab.$$

Example 12.2: Proof

Proof: Consider the difference of $a^2 + b^2 + \frac{1}{b^2} + 2$ and $2 \cdot \frac{a}{b} + 2ab$. So

$$a^2 + b^2 + \frac{1}{b^2} + 2 - \left(2 \cdot \frac{a}{b} + 2ab\right) =$$

$$= a^2 + \left(b + \frac{1}{b}\right)^2 - 2a\left(b + \frac{1}{b}\right) = \left(b + \frac{1}{b} - a\right)^2.$$

As $\left(b + \frac{1}{b} - a\right)^2 \geq 0$, then $a^2 + b^2 + \frac{1}{b^2} + 2 \geq 2 \cdot \frac{a}{b} + 2ab$.

Example 12.3:

Prove that
$$a^6 - 4a^4 + 2a^3 + 4a^2 - 4a + 1 \geq 0.$$

Example 12.3: Proof

Proof: Note that

$$a^6 - 4a^4 + 2a^3 + 4a^2 - 4a + 1 = (a^3 + 1)^2 - 4a^4 + 4a^2 - 4a =$$

$$= (a^3 + 1)^2 - 4a(a^3 + 1) + 4a^2 = (a^3 + 1 - 2a)^2.$$

As $(a^3 + 1 - 2a)^2 \geq 0$, then

$$a^6 - 4a^4 + 2a^3 + 4a^2 - 4a + 1 \geq 0.$$

12.1 Practice Problems

Prove the following inequalities:

(a) $\frac{a+b}{2} \geq \sqrt{ab}$, where $a \geq 0, b \geq 0$.

(b) $\frac{a^2+b^2}{2} \geq (\frac{a+b}{2})^2$.

(c) (a) $\frac{a}{b} + \frac{b}{a} \geq 2$, where $ab > 0$.

  (b) $\frac{a}{b} + \frac{b}{a} \leq -2$, where $ab < 0$.

(d) $\sqrt{ab} \geq \frac{2}{\frac{1}{a}+\frac{1}{b}}$, where $a > 0, b > 0$.

(e) $\frac{a+b}{2} \geq \frac{2}{\frac{1}{a}+\frac{1}{b}}$, where $a > 0, b > 0$.

(f) $\sqrt{\frac{a^2+b^2}{2}} \geq \frac{a+b}{2}$.

(g) $(a^2+b^2)(x^2+y^2) \geq (ax+by)^2$.

(h) $((a_1)^2 + (a_2)^2 + (a_3)^2)((b_1)^2 + (b_2)^2 + (b_3)^2) \geq$

$\geq (a_1 b_2 - a_2 b_1)^2 + (a_3 b_2 - a_2 b_3)^2 + (a_3 b_1 - a_1 b_3)^2$.

(i) $\frac{a^2}{c} + \frac{b^2}{d} \geq \frac{(a+b)^2}{c+d}$, where $c > 0, d > 0$.

(j) $\frac{x^2}{(x-1)^2} + \frac{y^2}{(y-1)^2} + \frac{z^2}{(z-1)^2} \geq 1$, where $xyz = 1$.

12.1.1 Proofs

(1) We have $\frac{a+b}{2} - \sqrt{ab} = (\frac{\sqrt{a}-\sqrt{b}}{\sqrt{2}})^2 \geq 0$. Thus $\frac{a+b}{2} \geq \sqrt{ab}$.

(2) We have $\frac{a^2+b^2}{2} - (\frac{a+b}{2})^2 = (\frac{a-b}{2})^2 \geq 0$. Thus $\frac{a^2+b^2}{2} \geq (\frac{a+b}{2})^2$.

(3) (a) We have $\frac{a}{b} + \frac{b}{a} - 2 = (\sqrt{\frac{a}{b}} - \sqrt{\frac{b}{a}})^2$. Thus $\frac{a}{b} + \frac{b}{a} \geq 2$.

(b) We have $-2 - (\frac{a}{b} + \frac{b}{a}) = (\sqrt{-\frac{a}{b}} - \sqrt{-\frac{b}{a}})^2$. Thus $-2 \geq \frac{a}{b} + \frac{b}{a}$.

(4) We have $\sqrt{ab} - \frac{2}{\frac{1}{a}+\frac{1}{b}} = \left(\sqrt{\sqrt{ab}} \cdot \frac{\sqrt{a}-\sqrt{b}}{\sqrt{a+b}}\right)^2 \geq 0$. Thus $\sqrt{ab} \geq \frac{2}{\frac{1}{a}+\frac{1}{b}}$.

(5) We have $\frac{a+b}{2} - \frac{2}{\frac{1}{a}+\frac{1}{b}} = \left(\frac{a-b}{\sqrt{2(a+b)}}\right)^2 \geq 0$. Thus $\frac{a+b}{2} \geq \frac{2}{\frac{1}{a}+\frac{1}{b}}$.

(6) When $a^2 + b^2 > 0$, we have

$$\sqrt{\frac{a^2+b^2}{2}} - \left|\frac{a+b}{2}\right| = \frac{\frac{a^2+b^2}{2} - \left|\frac{a+b}{2}\right|^2}{\sqrt{\frac{a^2+b^2}{2}} + \left|\frac{a+b}{2}\right|} = \left(\frac{a-b}{2\sqrt{\sqrt{\frac{a^2+b^2}{2}} + \left|\frac{a+b}{2}\right|}}\right)^2 \geq 0.$$

Thus $\sqrt{\frac{a^2+b^2}{2}} \geq \left|\frac{a+b}{2}\right|$. We have $\left|\frac{a+b}{2}\right| \geq \frac{a+b}{2}$.

From these two inequalities we get that $\sqrt{\frac{a^2+b^2}{2}} \geq \frac{a+b}{2}$.

(7) We have $(a^2 + b^2)(x^2 + y^2) - (ax + by)^2 = (ay - bx)^2 \geq 0$.
Thus $(a^2 + b^2)(x^2 + y^2) \geq (ax + by)^2$.

(8) We have $((a_1)^2 + (a_2)^2 + (a_3)^2)((b_1)^2 + (b_2)^2 + (b_3)^2) - (a_1b_2 - a_2b_1)^2 - (a_3b_2 - a_2b_3)^2 - (a_3b_1 - a_1b_3)^2 = (a_1b_1 + a_2b_2 + a_3b_3)^2 \geq 0$.

Thus $((a_1)^2 + (a_2)^2 + (a_3)^2)((b_1)^2 + (b_2)^2 + (b_3)^2) \geq$
$\geq (a_1b_2 - a_2b_1)^2 + (a_3b_2 - a_2b_3)^2 + (a_3b_1 - a_1b_3)^2$.

(9) We have
$$\frac{a^2}{c}+\frac{b^2}{d}-\frac{(a+b)^2}{c+d}=\left(\frac{ad-bc}{\sqrt{cd(c+d)}}\right)^2\geq 0.$$

Thus $\frac{a^2}{c}+\frac{b^2}{d}\geq\frac{(a+b)^2}{c+d}$.

(10) Note that, if we prove that
$$\frac{x^2}{(x-1)^2}+\frac{y^2}{(y-1)^2}+\frac{z^2}{(z-1)^2}-1=\left(\frac{x}{x-1}+\frac{y}{y-1}+\frac{z}{z-1}-1\right)^2\geq 0.$$

Then, we get that
$$\frac{x^2}{(x-1)^2}+\frac{y^2}{(y-1)^2}+\frac{z^2}{(z-1)^2}\geq 1.$$

Now, let us prove that
$$\left(\frac{x}{x-1}+\frac{y}{y-1}+\frac{z}{z-1}-1\right)^2=\left(\left(\frac{x}{x-1}+\frac{y}{y-1}\right)+\left(\frac{z}{z-1}-1\right)\right)^2=$$
$$=\left(\frac{x}{x-1}+\frac{y}{y-1}\right)^2+2\left(\frac{x}{x-1}+\frac{y}{y-1}\right)\left(\frac{z}{z-1}-1\right)+\left(\frac{z}{z-1}-1\right)^2=$$
$$=\frac{x^2}{(x-1)^2}+\frac{y^2}{(y-1)^2}+\frac{z^2}{(z-1)^2}+2\cdot\frac{xy(x-1)+yx(x-1)+xx(y-1)}{(x-1)(y-1)(z-1)}+$$
$$+\frac{-x(y-1)(z-1)-y(x-1)(z-1)-z(x-1)(y-1)}{(x-1)(y-1)(z-1)}+$$
$$+\frac{(x-1)(y-1)(z-1)}{(x-1)(y-1)(z-1)}-1=\frac{x^2}{(x-1)^2}+\frac{y^2}{(y-1)^2}+\frac{z^2}{(z-1)^2}-1.$$

(b) **Sum of nonnegative squares**
To solve some inequalities we may use the fact that the sum of squares of some expressions is nonnegative.

Example 12.4:
Prove the following inequality.
$$a^2 + b^2 + c^2 \geq ab + bc + ac.$$

Example 12.4: Proof
Proof: We have $a^2 + b^2 + c^2 - (ab + bc + ac) = \frac{2a^2+2b^2+2c^2-2ab-2bc-2ac}{2} = \frac{(a-b)^2+(b-c)^2+(c-a)^2}{2} = (\frac{a-b}{\sqrt{2}})^2 + (\frac{b-c}{\sqrt{2}})^2 + (\frac{c-a}{\sqrt{2}})^2 \geq 0.$

Thus $a^2 + b^2 + c^2 \geq ab + bc + ac$.

Example 12.5:
Prove the following inequality.
$$((a_1)^2 + (a_2)^2 + (a_3)^2)((b_1)^2 + (b_2)^2 + (b_3)^2) \geq (a_1 b_1 + a_2 b_2 + a_3 b_3)^2.$$

Example 12.5: Proof
Proof: We have (see the solution of problem 12.1.8) $((a_1)^2 + (a_2)^2 + (a_3)^2)((b_1)^2 + (b_2)^2 + (b_3)^2) - (a_1 b_1 + a_2 b_2 + a_3 b_3)^2 = (a_1 b_2 - a_2 b_1)^2 + (a_2 b_3 - a_3 b_2)^2 + (a_1 b_3 - a_3 b_1)^2 \geq 0.$

Thus $((a_1)^2 + (a_2)^2 + (a_3)^2)((b_1)^2 + (b_2)^2 + (b_3)^2) \geq (a_1 b_1 + a_2 b_2 + a_3 b_3)^2$.

12.2 Practice Problems

Prove the following inequalities:

(1) $(a+b+c)(\frac{1}{a}+\frac{1}{b}+\frac{1}{c}) \geq 9$, where $a > 0, b > 0, c > 0$.

(2) $a+b+\frac{9}{a-1}+\frac{16}{b-4} \geq 19$, where $a > 1, b > 4$.

(3) $\frac{ab}{c}+\frac{bc}{a}+\frac{ac}{b} \geq |a|+|b|+|c|$, where $abc > 0$.

(4) $x^3(x-y)^3 + y^3(y-z)^3 + z^3(z-x)^3 \geq 3xyz(x-y)(y-z)(z-x)$.

(5) $\frac{(a_1)^2}{b_1}+\frac{(a_2)^2}{b_2}+\frac{(a_3)^2}{b_3} \geq \frac{(a_1+a_2+a_3)^2}{b_1+b_2+b_3}$, where $b_1 > 0, b_2 > 0, b_3 > 0$.

(6) $\frac{a}{b}+\frac{b}{c}+\frac{c}{a} \geq 3$, where $a > 0, b > 0, c > 0$.

(7) $(2x^3+1)(2y^3+1)(2z^3+1) \geq 27$, where $x > 0, y > 0, z > 0$ and $xyz = 1$.

(8) $x+y+z \geq 3$, where $x > 0, y > 0, z > 0$ and $xyz = 1$.

(9) $(ab+1)^2 + (a-10)^2 + (b+10)^2 \geq 99$.

(10) $\frac{a^2+b^2+c^2+d^2+e^2}{5} \geq (\frac{a+b+c+d+e}{5})^2$.

12.2.1 Proofs

(1) We have

$$(a+b+c)\left(\frac{1}{a}+\frac{1}{b}+\frac{1}{c}\right) - 9 = \frac{a}{b} - 2 + \frac{b}{a} + \frac{b}{c} - 2 + \frac{b}{c} + \frac{a}{c} - 2 + \frac{c}{a} =$$

$$\left(\sqrt{\frac{a}{b}} - \sqrt{\frac{b}{a}}\right)^2 + \left(\sqrt{\frac{b}{c}} - \sqrt{\frac{c}{b}}\right)^2 + \left(\sqrt{\frac{a}{c}} - \sqrt{\frac{c}{a}}\right)^2 \geq 0.$$

Thus, it follows that

$$(a+b+c)\left(\frac{1}{a}+\frac{1}{b}+\frac{1}{c}\right) \geq 9.$$

Alternative proof. For two terms AM-GM (arithmetic mean - geometric mean) inequality is $\frac{x+y}{2} \geq \sqrt{xy}$. Here, we need to use AM-GM inequality for three terms, which is $\frac{x+y+z}{3} \geq \sqrt[3]{xyz}$, then

$$(a+b+c)\left(\frac{1}{a}+\frac{1}{b}+\frac{1}{c}\right) \geq 3\sqrt[3]{abc} \cdot 3\sqrt[3]{\frac{1}{a} \cdot \frac{1}{b} \cdot \frac{1}{c}} = 9.$$

(2) Use the identity $x^2 - 2xy + y^2 = (x-y)^2$, then

$$a + b + \frac{9}{a-1} + \frac{16}{b-4} - 19 = a - 1 - 6 + \frac{9}{a-1} + b - 4 - 8 + \frac{16}{b-4} =$$

$$= \left(\sqrt{a-1} - \frac{3}{\sqrt{a-1}}\right)^2 + \left(\sqrt{b-4} - \frac{4}{\sqrt{b-4}}\right)^2 \geq 0.$$

Thus, it follows that

$$a + b + \frac{9}{a-1} + \frac{16}{b-4} \geq 19.$$

Alternative proof. Let us use the inequality from example 12.1, we will write it as $x^2 + y^2 \geq 2xy$. Note that

$$a + b + \frac{9}{a-1} + \frac{16}{b-4} - 19 = a - 1 + \frac{9}{a-1} + b - 4 + \frac{16}{b-4} - 14.$$

Take $x^2 = a - 1$ and $y^2 = \frac{9}{a-1}$, then

$$x^2 + y^2 \geq 2xy = 2\sqrt{a-1} \cdot \sqrt{\frac{9}{a-1}} = 2 \cdot 3 = 6.$$

In a similar way, if we take $x^2 = b - 4$ and $y^2 = \frac{16}{b-4}$, we get that

$$b - 4 + \frac{16}{b-4} \geq 2\sqrt{b-4} \cdot \sqrt{\frac{16}{b-4}} = 2 \cdot 4 = 8.$$

So, we get

$$a + b + \frac{9}{a-1} + \frac{16}{b-4} - 19 = a - 1 + \frac{9}{a-1} + b - 4 + \frac{16}{b-4} - 14 \geq 6 + 8 - 14 = 0.$$

This ends the proof.

(3) Note that

$$\frac{ab}{c} + \frac{bc}{a} + \frac{ac}{b} - |a| - |b| - |c| = \frac{1}{2} \cdot \frac{ab}{c} - |a| + \frac{1}{2} \cdot \frac{ac}{b} + \frac{1}{2} \cdot \frac{ab}{c} - |b| + \frac{1}{2} \cdot \frac{bc}{a} + \frac{1}{2} \cdot \frac{bc}{a} -$$

$$-|c| + \frac{1}{2} \cdot \frac{ac}{b} = \left(\sqrt{\frac{ab}{2c}} - \sqrt{\frac{ac}{2b}}\right)^2 + \left(\sqrt{\frac{ab}{2c}} - \sqrt{\frac{bc}{2a}}\right)^2 + \left(\sqrt{\frac{bc}{2a}} - \sqrt{\frac{ac}{2b}}\right)^2 \geq 0.$$

Thus $\frac{ab}{c} + \frac{bc}{a} + \frac{ac}{b} \geq |a| + |b| + |c|$.

(4) Let us use the following identity:

$$a^3 + b^3 + c^3 - 3abc = (a + b + c)(a^2 + b^2 + c^2 - ab - bc - ac).$$

One can easily prove this identity by opening the parenthesis and cancelling all terms that are the same.

The identity is considered to be important, since it is possible to solve third degree equations using that.

According to the example 4, we have

$$a^2 + b^2 + c^2 - ab - bc - ac \geq 0.$$

Thus of $a+b+c \geq 0$, then we have
$$a^3 + b^3 + c^3 \geq 3abc.$$

From the example 4, we have $x(x-y) + y(y-z) + z(z-x) \geq 0$.

Thus, we have that
$$(x(x-y))^3 + (y(y-z))^3 + (z(z-x))^3 \geq 3x(x-y)y(y-x)z(z-x).$$

(5) Note that
$$\frac{a^2}{c} + \frac{b^2}{d} - \frac{(a+b)^2}{c+d} = \left(\frac{ad-bc}{\sqrt{cd(c+d)}}\right)^2,$$

where $c > 0, d > 0$, using this we have
$$\frac{a_1^2}{b_1} + \frac{a_2^2}{b_2} + \frac{a_3^2}{b_3} - \frac{(a_1+a_2+a_3)^2}{b_1+b_2+b_2} =$$
$$\frac{a_1^2}{b_1} + \frac{a_2^2}{b_2} - \frac{(a_1+a_2)^2}{b_1+b_2} + \frac{(a_1+a_2)^2}{b_1+b_2} + \frac{a_3^2}{b_3} - \frac{(a_1+a_2+a_3)^2}{b_1+b_2+b_2} =$$
$$\left(\frac{a_1 b_2 - a_2 b_1}{\sqrt{b_1 b_2 (b_1+b_2)}}\right)^2 + \left(\frac{(a_1+a_2)b_3 - (b_1+b_2)a_3}{\sqrt{(b_1+b_2)b_3(b_1+b_2+b_3)}}\right)^2 \geq 0$$

Thus $\frac{a_1^2}{b_1} + \frac{a_2^2}{b_2} + \frac{a_3^2}{b_3} \geq \frac{(a_1+a_2+a_3)^2}{b_1+b_2+b_2}$.

(6) $\frac{a}{b} + \frac{b}{c} + \frac{c}{a} - 3 = \left(\sqrt{\frac{a}{b}} - \sqrt{\frac{b}{c}}\right)^2 + \left(\sqrt{2\sqrt{\frac{a}{c}}} - \sqrt{2\sqrt{\frac{c}{a}}}\right)^2 + \left(\sqrt{\frac{c}{a}} - 1\right)^2 \geq 0.$

Thus $\frac{a}{b} + \frac{b}{c} + \frac{c}{a} \geq 3$

Alternative proof. We have
$$\frac{a}{b} + \frac{b}{c} + \frac{c}{a} = \frac{a \cdot a}{b \cdot a} + \frac{b \cdot b}{c \cdot b} + \frac{c \cdot c}{a \cdot c} = \frac{a^2}{ab} + \frac{b^2}{bc} + \frac{c^2}{ab} \geq \frac{(a+b+c)^2}{ab+bc+ac},$$

according to problem 12.2.5.

On the other hand, according to the example 4, we have

$$\frac{(a+b+c)^2}{ab+bc+ac} = \frac{a^2+b^2+c^2+2(ab+bc+ac)}{ab+bc+ac} = 2+\frac{a^2+b^2+c^2}{ab+bc+ac} \geq 2+1.$$

Thus $\frac{(a+b+c)^2}{ab+bc+ac} \geq 3$.

Combining all these inequalities, we get

$$\frac{a}{b} + \frac{b}{c} + \frac{c}{a} \geq 3.$$

(7) According to the solution of problem 12.2.4, we have

$$2x^3 + 1 = x^3 + x^3 - 1^3 - 3 \cdot x \cdot x \cdot 1 + 3x^2 \geq 0 + 3x^2 = 3x^2.$$

Similarly, we get $2y^3 + 1 \geq 3y^2$ and $2z^3 + 1 \geq 3z^2$.

Multiplying these three inequalities we get

$$(2x^3+1)(2y^3+1)(2z^3+1) \geq 3x^2 \cdot 3y^2 \cdot 3z^2.$$

On the other hand, we have

$$3x^2 \cdot 3y^2 \cdot 3z^2 = 27(xyz)^2 = 27.$$

Thus $(2x^3+1)(2y^3+1)(2z^3+1) \geq 27$.

(8) We have $x + y + z = x + y + \frac{1}{xy} = \frac{x}{1} + \frac{1}{\frac{1}{y}} + \frac{\frac{1}{y}}{x}$.

According to problem 12.2.6, we have $\frac{x}{1} + \frac{1}{\frac{1}{y}} + \frac{\frac{1}{y}}{x} \geq 3$.

Thus $x + y + z \geq 3$.

(9) Note that

$$(ab+1)^2 + (a-10)^2 + (b+10)^2 = (ab+1)^2 + (a-b-10)^2 + 99 \geq 99.$$

(10) Note that

$$\frac{a^2+b^2+c^2+d^2+e^2}{5} - \left(\frac{a+b+c+d+e}{5}\right)^2 =$$

$$= \left(\frac{1}{\sqrt{5}}\left(a-\left(\frac{a+b+c+d+e}{5}\right)^2\right)\right)^2 + \left(\frac{1}{\sqrt{5}}\left(b-\left(\frac{a+b+c+d+e}{5}\right)^2\right)\right)^2 +$$

$$+ \left(\frac{1}{\sqrt{5}}\left(c-\left(\frac{a+b+c+d+e}{5}\right)^2\right)\right)^2 + \left(\frac{1}{\sqrt{5}}\left(d-\left(\frac{a+b+c+d+e}{5}\right)^2\right)\right)^2 +$$

$$+ \left(\frac{1}{\sqrt{5}}\left(e-\left(\frac{a+b+c+d+e}{5}\right)^2\right)\right)^2 \geq 0.$$

Thus, we get that

$$\frac{a^2+b^2+c^2+d^2+e^2}{5} \geq \left(\frac{a+b+c+d+e}{5}\right)^2.$$

(c) **Evaluate the Addends**
To prove some inequalities we need to evaluate the addends, in case of product we need to evaluate the multipliers.

> **Example 12.6:**
>
> Let $a \geq 0, b \geq 0$. Prove the following inequality
> $$\frac{a+b}{1+a+b} \leq \frac{a}{1+a} + \frac{b}{1+b}.$$

> **Example 12.6: Proof**
> **Proof:** Note that $1 + a \leq 1 + a + b$, thus $\frac{a}{1+a} \geq \frac{a}{1+a+b}$. Similarly, we have that $\frac{b}{1+b} \geq \frac{a}{1+a+b}$. Adding the last two inequalities we get the inequality that we needed to prove.

> **Example 12.7:**
>
> Prove the following inequality.
> $$1 + \frac{1}{2^3} + \cdots + \frac{1}{2022^3} < 1.25.$$

> **Example 12.7: Proof**
> **Proof:** Note that
> $$1 + \cdots + \frac{1}{2^3} + \frac{1}{2022^3} = 1 + \frac{2-1}{2^3} + \frac{3-2}{3^3} + \frac{2022-2021}{2022^3} =$$
> $$1.25 - \frac{1}{2^3} + \frac{1}{3^2} - \frac{2}{3^3} + \cdots + \frac{1}{2022^2} - \frac{2021}{2022^3} =$$
> $$1.25 - \left(\frac{1}{2^3} - \frac{1}{3^2}\right) - \left(\frac{2}{3^3} - \frac{1}{4^2}\right) - \cdots - \left(\frac{2020}{2021^3} - \frac{1}{2022^2}\right) - \frac{2021}{2022^3} < 1.25.$$
> Since if k is a natural number then $\frac{k}{(k+1)^3} > \frac{1}{(k+2)^2}$.
>
> Indeed, we have $\frac{k}{(k+1)^3} - \frac{1}{(k+2)^2} = \frac{k^2+k-1}{(k+1)^3(k+2)^2} > 0$. Thus $\frac{k}{(k+1)^3} > \frac{1}{(k+2)^2}$.

12.3 Practice Problems

Prove the following inequalities.

(1) $\frac{1}{501} + \frac{1}{502} + \cdots + \frac{1}{1000} > \frac{3}{5}$.

(2) $\frac{1}{2} + \frac{1}{3} - \frac{1}{4} + \frac{1}{5} + \cdots + \frac{1}{998} - \frac{1}{999} + \frac{1}{1000} < \frac{2}{5}$.

(3) $\frac{1}{2^2} + \frac{1}{3^2} + \cdots + \frac{1}{(1000)^2} < \frac{3}{4}$.

(4) $\frac{\sqrt{2}}{20} < \frac{1}{2} \cdot \frac{3}{4} \cdots \cdots \frac{99}{100} < \frac{\sqrt{3}}{20}$.

(5) Let $a > 0, b > 0, c > 0, d > 0$ prove that

$$1 < \frac{a}{a+b+d} + \frac{b}{a+b+c} + \frac{c}{b+c+d} + \frac{d}{a+c+d} < 2.$$

(6) $\frac{a-b}{a+b} + \frac{b-c}{b+c} + \frac{c-a}{c+a} < 1$, where $a > 0, b > 0, c > 0$.

(7) $(\frac{a-b}{a+b})^{2023} + (\frac{b-c}{b+c})^{2023} + (\frac{c-a}{c+a})^{2023} < 1$, where $a > 0, b > 0, c > 0$.

(8) Let $0 \leq a \leq 1, 0 \leq b \leq 1, 0 \leq c \leq 1$, prove that

$$\frac{a}{b+c+1} + \frac{b}{c+a+1} + \frac{c}{a+b+1} + (1-a)(1-b)(1-c) \leq 1.$$

(9) Let $0 < a \leq \frac{1}{2}, 0 < b \leq \frac{1}{2}, 0 < c \leq \frac{1}{2}$ and let $\bar{a} = 1-a, \bar{b} = 1-b, \bar{c} = 1-c$, prove that

$$\frac{a^3 + b^3 + c^3}{abc} \geq \frac{\bar{a}^3 + \bar{b}^3 + \bar{c}^3}{\bar{a} \cdot \bar{b} \cdot \bar{c}}.$$

(10) $\frac{a^4}{b^3} + \frac{b^4}{c^3} + \frac{c^4}{a^3} \geq \frac{a^3}{b^2} + \frac{b^3}{c^2} + \frac{c^3}{a^2}$, where $a > 0, b > 0, c > 0$.

12.3.1 Proofs

(1) We have

$$\frac{1}{501}+\cdots+\frac{1}{1000} = (\frac{1}{501}+\cdots+\frac{1}{700})+(\frac{1}{701}+\cdots\frac{1}{900})+(\frac{1}{901}+\cdots+\frac{1}{1000}).$$

Note that the number of factions of $\frac{1}{501}+\frac{1}{502}+\cdots+\frac{1}{700}$ is 700, and the smallest fraction is $\frac{1}{700}$.

It follows that $\frac{1}{501}+\cdots+\frac{1}{700} > 200 \cdot \frac{1}{700}$, thus $\frac{1}{501}+\cdots+\frac{1}{700} > \frac{2}{7}$. Similarly, we get $\frac{1}{701}+\cdots+\frac{1}{900} > \frac{2}{9}$ and $\frac{1}{901}+\cdots+\frac{1}{1000} > \frac{1}{10}$.

Thus $\frac{1}{501}+\cdots+\frac{1}{1000} > \frac{2}{7}+\frac{2}{9}+\frac{1}{10} = \frac{32}{63}+\frac{1}{10} > \frac{1}{2}+\frac{1}{10} = \frac{3}{5}$.
Thus

$$\frac{1}{501}+\cdots+\frac{1}{1000} > \frac{3}{5}.$$

(2) We have

$$\frac{1}{2}-\frac{1}{3}+\frac{1}{4}-\frac{1}{5}+\cdots+\frac{1}{998}-\frac{1}{999}+\frac{1}{1000} =$$
$$= 2(\frac{1}{2}+\frac{1}{4}+\cdots+\frac{1}{1000}) - (\frac{1}{2}+\frac{1}{3}+\cdots+\cdot+\frac{1}{1000}) =$$
$$= 1+\frac{1}{2}+\frac{1}{3}+\cdots+\frac{1}{500} - (\frac{1}{2}+\frac{1}{3}+\cdots+\frac{1}{1000}) =$$
$$= 1 - (\frac{1}{501}+\cdots+\frac{1}{1000}).$$

According to problem 12.3.1, we have that $\frac{1}{501}+\cdots+\frac{1}{1000} > \frac{3}{5}$.

Thus $\frac{1}{2}-\frac{1}{3}+\frac{1}{4}-\frac{1}{5}+\cdots \frac{1}{998}-\frac{1}{999}+\frac{1}{1000} = 1-(\frac{1}{501}+\cdots+\frac{1}{1000}) < 1-\frac{3}{5} = \frac{2}{5}$.

(3) We have $3^2 > 2 \cdot 3, 4^2 . 3 \cdot 4, \cdot, 999^2 > 998 \cdot 999$.

Thus
$$\frac{1}{2^2} + \frac{1}{3^2} + \cdots + \frac{1}{1000^2} < \frac{1}{2^2} + \frac{1}{2 \cdot 3} + \cdots + \frac{1}{998 \cdot 999} + \frac{1}{1000^2} =$$
$$\frac{1}{4} + \frac{1}{2} - \frac{1}{3} + \frac{1}{4} + \cdots + \frac{1}{998} - \frac{1}{999} + \frac{1}{1000^2} =$$
$$\frac{3}{4} - \left(\frac{1}{999} - \frac{1}{1000^2}\right) < \frac{3}{4}.$$

From this, we have that $\frac{1}{2^2} + \frac{1}{3^2} + \cdots + \frac{1}{1000^2} < \frac{3}{4}$.

(4) Prove that if k is a natural number then
$$\sqrt{\frac{k-1}{k}} < \frac{2k-1}{2k} \leq \sqrt{\frac{3k-2}{3k+1}}.$$

Indeed, we have $(2k-1)^2 k - 4k^2(k-1) = k > 0, thus (2k-1)^2 k > 4k^2(k-1)$, from which
$$\frac{2k-1}{2k} > \sqrt{\frac{k-1}{k}}.$$

We also have $(3k-2) \cdot 4k^2 - (2k-1)^2(3k+1) = k - 1 \geq 0$. Thus $(3k-2)4k^2 \geq (2k-1)^2(3k+1)$, from which
$$\frac{2k-1}{2k} \leq \sqrt{\frac{3k-2}{3k+1}}.$$

So we have that
$$\sqrt{\frac{1}{2}} < \frac{3}{4} < \sqrt{\frac{4}{7}}$$

$$\sqrt{\frac{2}{3}} < \frac{5}{6} < \sqrt{\frac{7}{10}}, \ldots,$$

$$\sqrt{\frac{49}{50}} < \frac{99}{100} < \sqrt{\frac{148}{151}}.$$

Multiplying all the inequalities we get:

$$\frac{1}{\sqrt{50}} < \frac{3}{4} \cdot \frac{5}{6} \cdot \ldots \cdot \frac{99}{100} < \sqrt{\frac{4}{151}}$$

from which,

$$\frac{1}{2\sqrt{50}} < \frac{1}{2} \cdot \frac{3}{4} \cdot \ldots \cdot \frac{99}{100} < \frac{1}{2}\sqrt{\frac{4}{151}}.$$

It is left to note that $\frac{1}{2\sqrt{50}} = \frac{\sqrt{2}}{20}$ and $\frac{1}{2}\sqrt{\frac{4}{151}} < \frac{\sqrt{3}}{20}$.

(5) We have that the denominators of the fractions $\frac{a}{a+b+d}, \frac{b}{a+b+c}, \frac{c}{b+c+d}, \frac{d}{a+c+d}$ is less than $a+b+c+d$. Thus

$$\frac{a}{a+b+d} + \frac{b}{a+b+c} + \frac{c}{b+c+d} + \frac{d}{a+c+d} >$$

$$\frac{a}{a+b+c+d} + \frac{b}{a+b+c+d} + \frac{c}{a+b+c+d} + \frac{d}{a+b+c+d} = 1.$$

Note that $\frac{a}{a+b+d} + \frac{b}{a+b+c} < \frac{a}{a+b} + \frac{b}{a+b} = 1$ and $\frac{c}{b+c+d} + \frac{d}{a+c+d} < \frac{c}{c+d} + \frac{d}{c+d} = 1$.

Adding the two inequalities we have

$$\frac{a}{a+b+d} + \frac{b}{a+b+c} + \frac{c}{b+c+d} + \frac{d}{a+c+d} < 2.$$

(6) We may consider that b and c are not bigger than a, in other case we write the inequality as $\frac{b-c}{b+c}+\frac{c-a}{c+a}+\frac{a-b}{a+b} < 1$ or $\frac{c-a}{c+a}+\frac{a-b}{a+b}+\frac{b-c}{b+c} < 1$.

We have $\frac{a-b}{a+b} < 1$, prove that $\frac{b-c}{b+c} \leq \frac{a-c}{c+a}$.

Indeed, $\frac{a-b}{a+b} - \frac{b-c}{b+c} = \frac{2c(a-b)}{(a+c)(b+c)} \geq 0$, thus $\frac{b-c}{b+c} \leq \frac{a-b}{a+b}$.

Adding the inequalities we get:
$$\frac{a-b}{a+b} + \frac{b-c}{b+c} + \frac{c-a}{c+a} < 1.$$

(7) We may consider that b and c are not bigger than a (see the solution of problem 12.3.6).

According to the solution of problem 12.3.6, we have
$$\frac{a-b}{a+b} < 1, \quad \frac{b-c}{b+c} \leq -\frac{c-a}{a+c}.$$

Thus
$$\left(\frac{a-b}{a+b}\right)^{2023} < 1^{2023}$$
and
$$\left(\frac{b-c}{b+c}\right)^{2023} \leq \left(-\frac{c-a}{c+a}\right)^{2023}.$$

From the last two equations we get:
$$\left(\frac{a-b}{a+b}\right)^{2023} + \left(\frac{b-c}{b+c}\right)^{2023} + \left(\frac{c-a}{c+a}\right)^{2023} < 1.$$

(8) We have that $(1-a)(1-b) \geq 0$, thus $a+b-1 \leq ab$ and

$$\frac{1}{a+b+1} - \left(1 - \frac{a+b}{2}\right) = \frac{(a+b)(a+b-1)}{2(1+a+b)} \leq \frac{a+b}{2(1+a+b)} \cdot ab \leq \frac{1}{3} \cdot ab.$$

We get that
$$\frac{1}{1+a+b} \leq 1 - \frac{a+b}{2} + \frac{ab}{3},$$

Thus
$$\frac{c}{a+b+1} \leq c - \frac{ac+bc}{2} + \frac{abc}{3}.$$

Similarly, we have that
$$\frac{b}{c+a+1} \leq b - \frac{ab+bc}{2} + \frac{abc}{3}$$

and
$$\frac{a}{b+c+1} \leq a - \frac{ab+ac}{2} + \frac{abc}{3}.$$

Summing the last three inequalities we get the inequality that needed to be proved.

(9) We have that

$$a^3 + b^3 + c^3 - 3abc = \frac{1}{2}(a+b+c)((a-b)^2 + (b-c)^2 + (c-a)^2),$$

thus

$$\frac{a^3+b^3+c^3}{abc} = 3 + \frac{1}{2}\left(\frac{1}{bc} + \frac{1}{ac} + \frac{1}{ab}\right)((a-b)^2 + (b-c)^2 + (c-a)^2).$$

Note that

$$\bar{a} \geq a, \bar{b} \geq b, \bar{c} \geq c, (\bar{a}-\bar{b})^2 = (a-b)^2, (\bar{b}-\bar{c})^2 = (b-c)^2, (\bar{c}-\bar{a})^2 = (c-a)^2.$$

Thus
$$\frac{1}{bc}+\frac{1}{ac}+\frac{1}{ab} \geq \frac{1}{\overline{b}\overline{c}}+\frac{1}{\overline{a}\overline{c}}+\frac{1}{\overline{a}\overline{b}},$$
and
$$\frac{a^3+b^3+c^3}{abc} \geq 3+\frac{1}{2}\left(\frac{1}{\overline{b}\overline{c}}+\frac{1}{\overline{a}\overline{c}}+\frac{1}{\overline{a}\overline{b}}\right)((a-b)^2+(b-c)^2+(c-a)^2) = \frac{\overline{a}^3+\overline{b}^3+\overline{c}^3}{\overline{a}\cdot\overline{b}\cdot\overline{c}}.$$

(10) Note that
$$\frac{a^4}{b^3}-\frac{a^3}{b^2} = \frac{a^3}{b^2}\left(\frac{a}{b}-1\right) \geq b\left(\frac{a}{b}-1\right),$$
as from one of the inequalities $\frac{a^3}{b^2} \geq b$ and $\frac{a}{b}-1 \geq 0$ follows the other one.

We get that
$$\frac{a^4}{b^3} \geq \frac{a^3}{b^2} + a - b,$$
Similarly, we have
$$\frac{b^4}{c^3} \geq \frac{b^3}{c^2} + b - c$$
and,
$$\frac{c^4}{a^3} \geq \frac{c^3}{a^2} + c - a$$
Thus
$$\frac{a^4}{b^3}+\frac{b^4}{c^3}+\frac{c^4}{a^3} \geq \frac{a^3}{b^2}+\frac{b^3}{c^2}+\frac{c^3}{a^2}.$$

Chapter 13

Geometry

(a) **Properties of the median to the hypotenuse of a right triangle.**

Consider right triangle ABC, where $\angle ACB = 90°$ (see the figure).

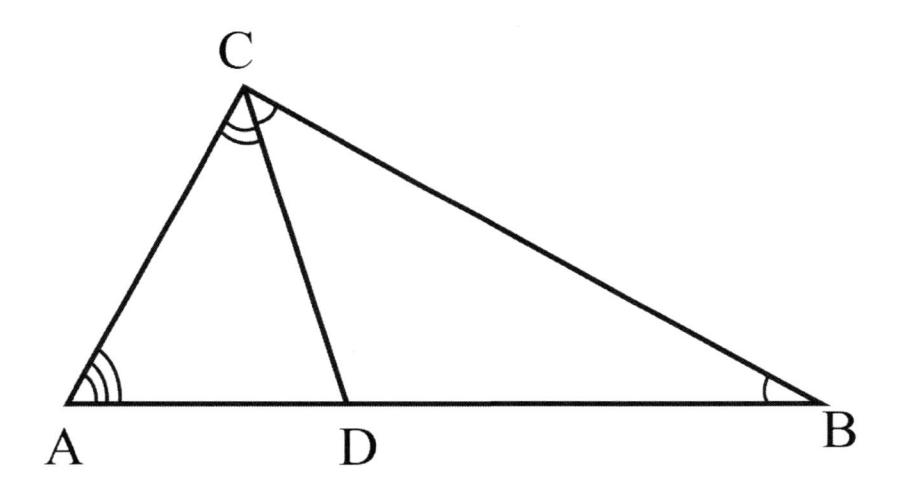

Choose a point D on hypotenuse AB so that $\angle DCB = \angle DBC$. We have $\angle DAC = 90° - \angle DBC = 90° - \angle DCB = \angle DCA$. Thus $\angle DAC = \angle DCA$.

Then $\angle DCB = \angle DBC$ and $\angle DAC = \angle DCA$, which means that triangles DBC and DAC are isosceles triangles.

We get the following theorem.

Theorem 13.1. *The median to the hypotenuse of a right triangle is equal to half of the hypotenuse.*

> **Example 13.1:**
>
> Suppose CH and C_1H_1 are the altitudes perpendicular to the hypotenuses of right triangles ABC and $A_1B_1C_1$, respectively. Given that $CH = C_1H_1$ and $AB = A_1B_1$. Prove that triangles ABC and $A_1B_1C_1$ are congruent.

Example 13.1: Proof
Proof: Let CM and C_1M_1 be the medians of right triangles ABC and $A_1B_1C_1$ (see the figure).

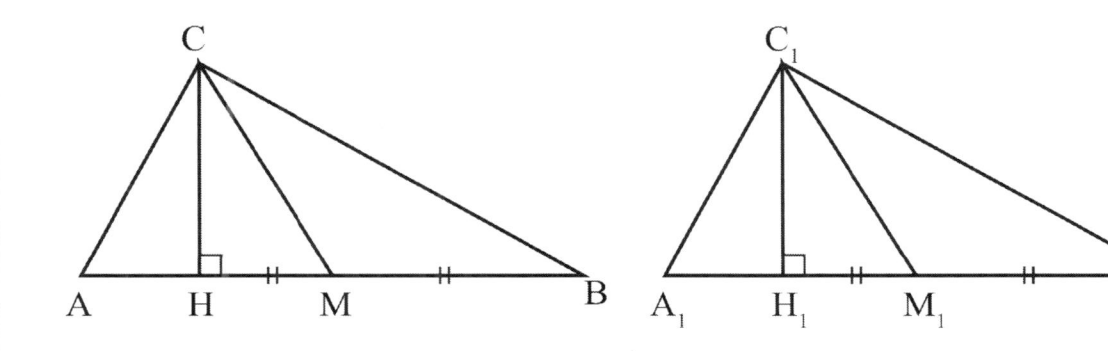

According to theorem 13.1, we have

$$CM = \frac{AB}{2} = \frac{A_1B_1}{2} = C_1M_1.$$

Thus, triangles CHM and $C_1H_1M_1$ are congruent ($CH = C_1H_1$ and $CM = C_1M_1$). We get that $HM = H_1M_1$, so

$$HB = HM + MB = H_1M_1 + \frac{AB}{2} = H_1M_1 + \frac{A_1B_1}{2} = H_1M_1 + M_1B_1 = H_1B_1.$$

So, $CH = C_1H_1$ and $HB = H_1B_1$. Then, right triangles CHB and $C_1H_1B_1$ are congruent. So, $CB = C_1B_1$.
According to the condition of the problem, we have $AB = A_1B_1$. Thus, right triangles ABC and $A_1B_1C_1$ are also congruent.

13.1 Practice Problems

Problem 13.1. Let ABC be a right triangle, so that $\angle ABC = 30°$, $\angle ACB = 90°$. Prove that $AC = \frac{AB}{2}$.

Problem 13.2. Let M be the midpoint of side AB of triangle ABC. Given that $2 \cdot CM = AB$. Find $\angle ACB$.

Problem 13.3. Let ABC be a right triangle, so that $\angle ABC = 30°$, $\angle ACB = 90°$. The perpendicular-bisector of AB intersects leg BC at point E. Let D be the midpoint of AB. Find $\frac{ED}{BC}$.

Problem 13.4. Let M and N be the midpoints of sides AB and BC of triangle ABC, respectively. Let BH be an altitude. The perimeter of triangle ABC is 60. Find the perimeter of triangle MNH.

Problem 13.5. Let ABC be a right triangle, so that $\angle ABC = 15°$, $\angle ACB = 90°$. Let CH be the altitude to the hypotenuse. Given that $AB = 20$. Find CH.

Problem 13.6. Let AA_1, BB_1, CC_1 be the altitudes and AA_2, BB_2, CC_2 be the medians of triangle ABC. Prove that the length of the broken line $A_2B_1C_2A_1B_2C_1A_2$ is equal to the perimeter of triangle ABC.

Problem 13.7. Let D and E be points outside triangle ABC. Given that $\angle ADB = \angle CEB = 90°$ and $\angle BAD = \angle BCE$. Let M be the midpoint of side AC. Prove that $DM = EM$.

Problem 13.8. Let AB be the greatest side of triangle ABC and $\angle ABC = 15°$. Let CH be an altitude and $4 \cdot CH = AB$. Find $\angle BAC$.

Problem 13.9. Let CD be the median of right triangle ABC to its hypotenuse. A line perpendicular to CD passes through point D and intersects AC at point E, it also intersects the extension of BC at point F. Let M be the midpoint of EF. Prove that $AB \perp CM$.

13.1.1 Solutions of Practice Problems

Problem 13.1: *Let ABC be a right triangle, so that $\angle ABC = 30°$, $\angle ACB = 90°$. Prove that $AC = \frac{AB}{2}$.*

Proof: Let CM be a median of triangle ABC (see the figure).

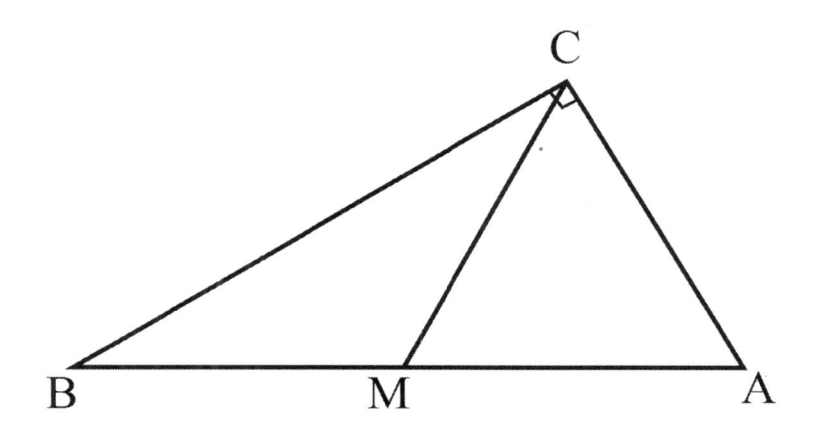

We have that $CM = AM$ and
$$\angle MAC = \angle BAC = 90° - \angle ABC = 60°.$$
Thus, it follows that
$$\angle MCA = \angle MAC = 60°,$$
and
$$\angle AMC = 180° - (\angle MCA + \angle MAC) = 60°.$$
We get that triangle AMC is an isosceles triangle. So
$$AC = AM = \frac{AB}{2}.$$

Problem 13.2: *Let M be the midpoint of side AB of triangle ABC. Given that $2 \cdot CM = AB$. Find $\angle ACB$.*

Answer: $90°$.

Solution: We have $2CM = AB = 2AM = 2BM$ (see the figure).

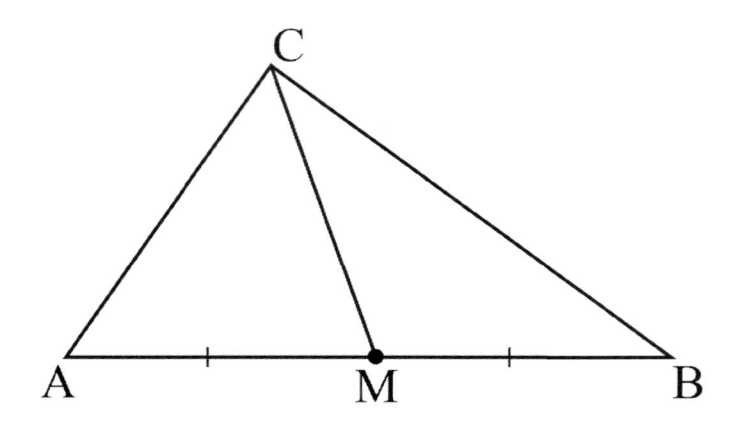

So, $CM = AM$ and $CM = BM$.

We get that triangles AMC and BMC are isosceles triangles. Then, $\angle MAC = \angle MCA$ and $\angle MBC = \angle MCB$.

Thus, it follows that

$$\angle ACB = \angle MCA + \angle MCB = \angle MAC + \angle MBC =$$

$$= \angle BAC + \angle ABC = 180° - \angle ACB.$$

We get that
$$ACB = 180° - \angle ACB = 90°.$$

Problem 13.3: *Let ABC be a right triangle, so that $\angle ABC = 30°$, $\angle ACB = 90°$. The perpendicular-bisector of AB intersects leg BC at point E. Let D be the midpoint of AB. Find $\frac{ED}{BC}$.*

Answer: $1:3$.

Solution: Let CD be a median of triangle ABC (see the figure).

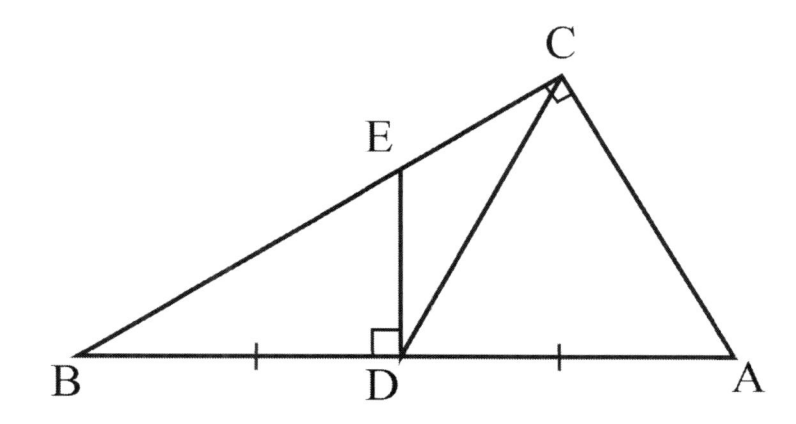

We have that $CD = BD$, thus
$$\angle BCD = \angle DBC = \angle ABC = 30°.$$

According to problem 13.1, we have that $BE = 2 \cdot ED$. Then,
$$\angle BDC = 180° - (\angle BCD + \angle DBC) = 120°,$$
and
$$\angle EDC = \angle BDC - 90° = 30° = \angle BCD.$$

We get that $ED = EC$. So, we have
$$BC = BE + EC = 2ED + ED = 3ED.$$

Thus, it follows that
$$\frac{ED}{BC} = \frac{1}{3}.$$

Problem 13.4: *Let M and N be the midpoints of sides AB and BC of triangle ABC, respectively. Let BH be an altitude. The perimeter of triangle ABC is 60. Find the perimeter of triangle MNH.*

Answer: 30.
Solution: Let us consider the following figure.

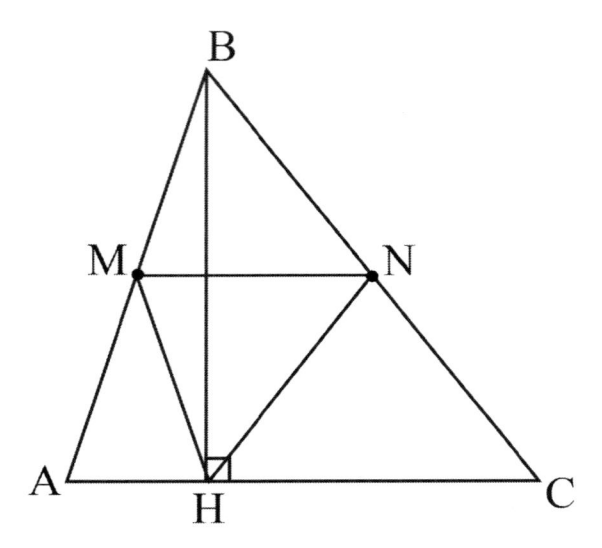

As HM and HN are medians of right triangles ABH and BHC, then $MH = \frac{AB}{2}$ and $NH = \frac{BC}{2}$.

Note that MN is a midsegment of triangle ABC, thus $MN = \frac{AC}{2}$. So

$$MH + NH + MN = \frac{AB}{2} + \frac{BC}{2} + \frac{AC}{2} = \frac{1}{2} \cdot 60 = 30.$$

Problem 13.5: *Let ABC be a right triangle, so that $\angle ABC = 15°$, $\angle ACB = 90°$. Let CH be the altitude to the hypotenuse. Given that $AB = 20$. Find CH.*

Answer: 5.
Solution: Let CM be a median of triangle ABC (see the figure).

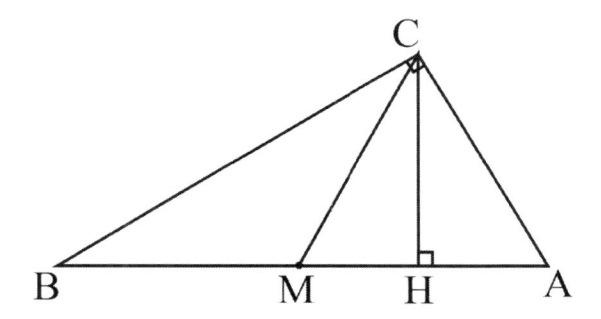

We have that $CM = BM$, thus $\angle MCB = \angle MBC = \angle ABC = 15°$. According to the exterior angle property, we have that $\angle HMC = \angle MCB + \angle MBC = 15° + 15° = 30°$ for triangle BMC.

So, from triangle MHC, we get $CH = \frac{CM}{2}$. Thus

$$CH = \frac{CM}{2} = \frac{AB}{4} = \frac{20}{4} = 5.$$

Problem 13.6: *Let AA_1, BB_1, CC_1 be the altitudes and AA_2, BB_2, CC_2 be the medians of triangle ABC. Prove that the length of the broken line $A_2B_1C_2A_1B_2C_1A_2$ is equal to the perimeter of triangle ABC.*

Solution: Consider the following figure.

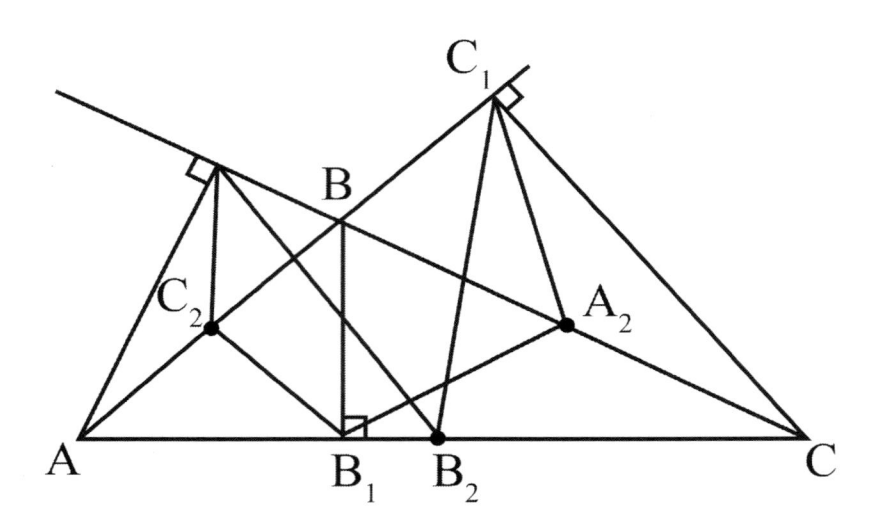

According to theorem 13.1, from right triangles BCB_1, ABB_1, ABA_1, AA_1C, ACC_1, BCC_1, we get that

$$A_2B_1 = \frac{BC}{2}, B_1C_2 = \frac{AB}{2}, C_2A_1 = \frac{AB}{2},$$

$$A_1B_2 = \frac{AC}{2}, B_2C_1 = \frac{AC}{2}, C_1A_2 = \frac{BC}{2}.$$

Thus, it follows that

$$A_2B_1 + B_1C_2 + C_2A_1 + A_1B_2 + B_2C_1 + C_1A_2 =$$

$$= 2 \cdot \frac{AB}{2} + 2 \cdot \frac{BC}{2} + 2 \cdot \frac{AC}{2} = AB + BC + AC.$$

Problem 13.7: *Let D and E be points outside triangle ABC. Given that $\angle ADB = \angle CEB = 90°$ and $\angle BAD = \angle BCE$. Let M be the midpoint of side AC. Prove that $DM = EM$.*

Solution: Let DN and EK be the medians of right triangles ABD and BEC (see the figure).

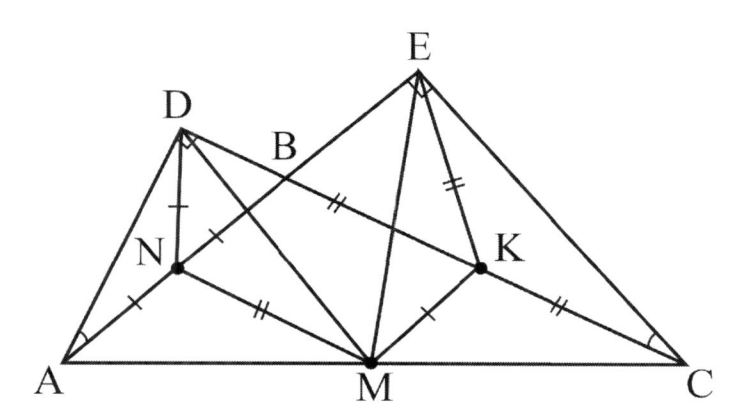

According to theorem 13.1, we have $DN = \frac{AB}{2}$ and $EK = \frac{BC}{2}$. As MN and MK are midsegments of triangle ABC, then $MN \| BC$, $MK \| AB$ and
$$MN = \frac{BC}{2}, MK = \frac{AB}{2}.$$
Thus, it follows that
$$DN = \frac{AB}{2} = MK, \quad EK = \frac{BC}{2} = MN.$$
Then, $\angle MNB = 180° - \angle ABC = \angle MKB$ and
$$\angle DNB = 2\angle BAD = 2\angle BCE = \angle EKB.$$
So, $\angle MNB + \angle DNB = \angle MKB + \angle EKB$. Then, $\angle MND = \angle MKE$.

By the side-angle-side (SAS) congruence postulate, triangles MND and EKM are congruent. Thus $DM = EM$.

Problem 13.8: *Let AB be the greatest side of triangle ABC and $\angle ABC = 15°$. Let CH be an altitude and $4 \cdot CH = AB$. Find $\angle BAC$.*

Answer: 75°.

Solution: Take a point D on side AB so that $\angle DCB = \angle ABC = 15°$ (see the figure).

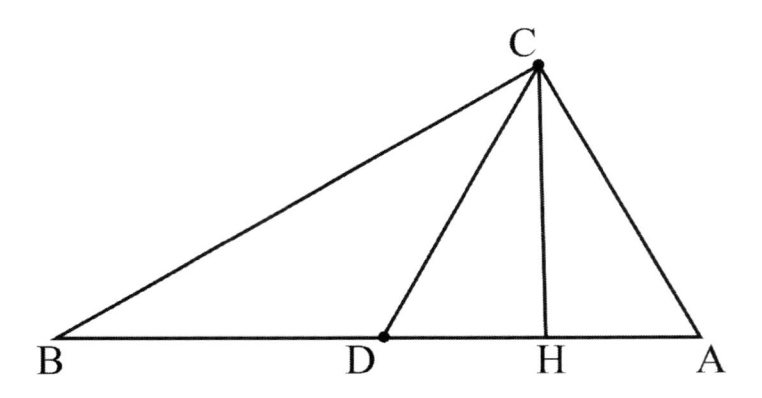

According to the property of an external angle, from BCD triangle we have $\angle CDH = 2 \cdot 15° = 30°$.

From triangle CDH, according to problem 13.1, we have $2CH = CD$. Thus, it follows that

$$2 \cdot CH = CD = BD.$$

According to the condition of the problem, we have $AB = 4CH$. So,

$$AD = AB - BD = 4 \cdot CH - 2 \cdot CH = 2 \cdot CH.$$

We get that
$$AD = BD = CD.$$

According to problem 13.5, we have $\angle ACB = 90°$. Thus $\angle BAC = 78°$.

Problem 13.9: *Let CD be the median of right triangle ABC to its hypotenuse. A line perpendicular to CD passes through point D and intersects AC at point E, it also intersects the extension of BC at point F. Let M be the midpoint of EF. Prove that $AB \perp CM$.*

Solution: Consider the following figure.

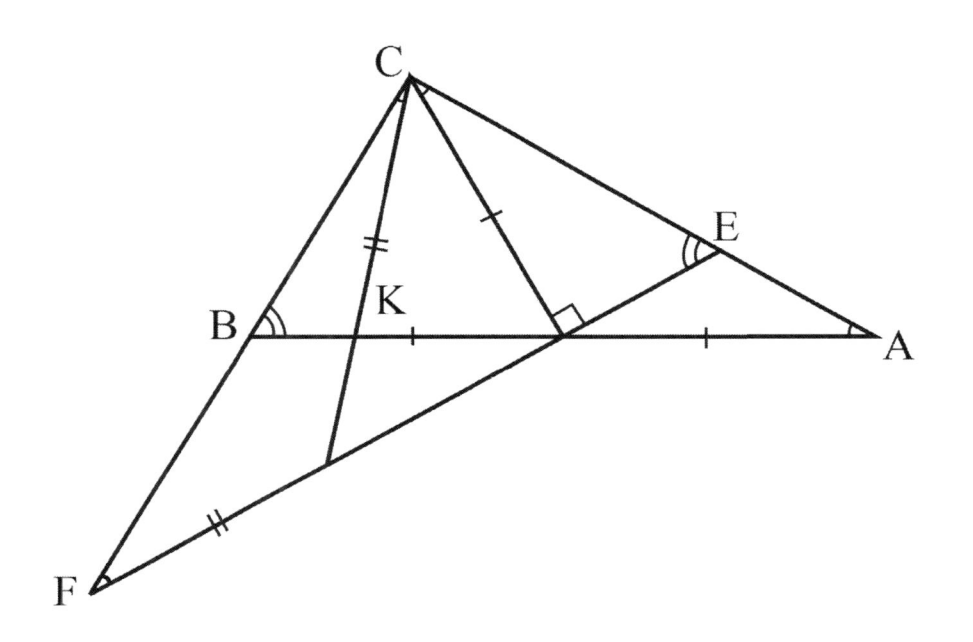

Let $\angle BAC = \alpha$, then from $CD = DA$ we have that $\angle DCA = \alpha$.

From triangle CDE, we have that $\angle CED = 90° - \alpha$ and from triangle CEF, we have that $\angle CFE = \alpha$.

As $CM = FM$, then $\angle MCB = \alpha$.
From triangle ABC, we have $\angle ABC = 90° - \alpha$.

From triangle BCK, we have $\angle BKC = 90°$.
This ends the proof.

(b) **Construct an isosceles triangle.**

The beauty of geometry lies also in the fact that, to solve problems, often one needs to construct additional shapes (for example triangles).

> **Example 13.2:**
> Let M and N be points on sides AB and AC of triangle ABC, respectively. Given that $\angle AC = 70°, \angle ACB = 50°, \angle MCB = 40°$ and $\angle NBC = 50°$. Find $\angle NMC$.

Example 13.2: Proof
Answer: $30°$.
Proof: Construct an isosceles triangle CKN (see the figure), then

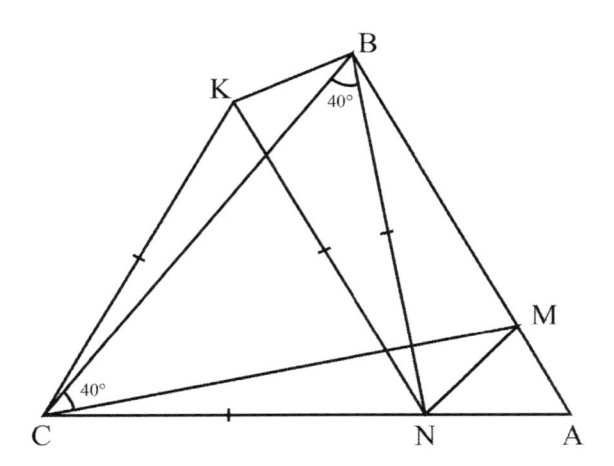

$\angle BCK = \angle NCK - \angle NCB = 60° - 50° = 10° = \angle NCB - \angle MCB = \angle NCM$. So $\angle NCM = \angle BCK$.
$\angle CMB = 180° - (\angle ABC + \angle MCB) = 180° - (70° + 40°) = 70° = \angle MBC$. Thus, $CB = CM$. As $CK = CM$, then $\angle NCM = \angle BCK$ and $CB = CM$. Then, triangles BCK and MCN are congruent. So

$$\angle NMC = \angle KBC = \angle KBN - \angle NBC = \frac{180° - \angle BNK}{2} - 50° =$$

$$\frac{180° - (\angle BNC - \angle KNC)}{2} - 50° = \frac{180° - (80° - 60°)}{2} - 50° = 30°.$$

Example 13.3:

Let BM be a median of triangle ABC. Given that $\angle ACB = 30°$ and $\angle MBC = 15°$. Find .

Example 13.3: Proof
Answer: $105°$.
Proof: Construct an equilateral triangle ANM (see the figure).

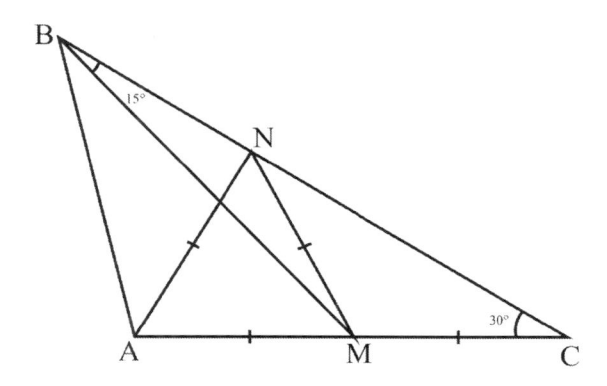

Note that the NMC triangle is equilateral and $\angle MNC = \angle MCN = \frac{1}{2}\angle AMN = \frac{1}{2} \cdot 60° = 30°$.

Thus the point N is on the BC side.

From BMN triangle we can get that $\angle NMB = \angle MNC - \angle MBN = 30° - 15° = 15°$.

Thus $BN = MN = AN$.

From this we can get that $\angle BAN = \frac{1}{2}\angle ANC = 45°$.

Thus, $\angle BAC = \angle BAN + \angle MAN = 45° + 60° = 105°$.

13.2 Practice Problems

Problem 13.10. M is a point inside an isosceles triangle ABC so that $\angle MBA = 10°$ and $BM = AC$. Let $\angle A = \angle C = 70°$. Find $\angle MAC$.

Problem 13.11. D is a point on side BC of an isosceles triangle ABC so that $BD = AC$ and $\angle A = \angle C = 80°$. Find $\angle BAD$.

Problem 13.12. M is a point inside a triangle ABC with the base AC so that $\angle MBA = 10°, \angle MBC = 30°$ and $BM = AC$. Find the angles of triangle AMC. (N.M. Sedrakyan)

Problem 13.13. M is a point inside an isosceles triangle ABC with the base BC so that $\angle MBC = 30°$ and $\angle MCB = 10°$. Let $\angle BAC = 80°$. Find $\angle AMC$.

Problem 13.14. M is a point inside a triangle ABC so that $\angle MAB = 20°$ and $\angle MBA = 10°$. Let $AB = AC$ and $\angle BAC = 100°$. Find $\angle AMC$.

Problem 13.15. $ABCD$ is a convex quadrilateral so that $\angle BCA = 20°$, $\angle BAC = \angle DBC = 30°$ and $\angle BDC = 70°$. Prove that $AD \parallel BC$.

Problem 13.16. Let equilateral triangles ACM, BDN be built on the diagonals AC, BD of a convex quadrilateral $ABCD$, so that points B, M are on the same side of line AC, and points C, N are on the same side of line BD. Let $MN = AB + CD$. Find $\angle BAD + \angle CDA$. (N.M. Sedrakyan)

13.2.1 Solutions of Practice Problems

Problem 13.10: *M is a point inside the isosceles triangle ABC so that $\angle MBA = 10°$ and $BM = AC$. Let $\angle A = \angle C = 70°$. Find $\angle MAC$.*

Answer: $50°$.

Solution: Let us build the ANC equilateral triangle.

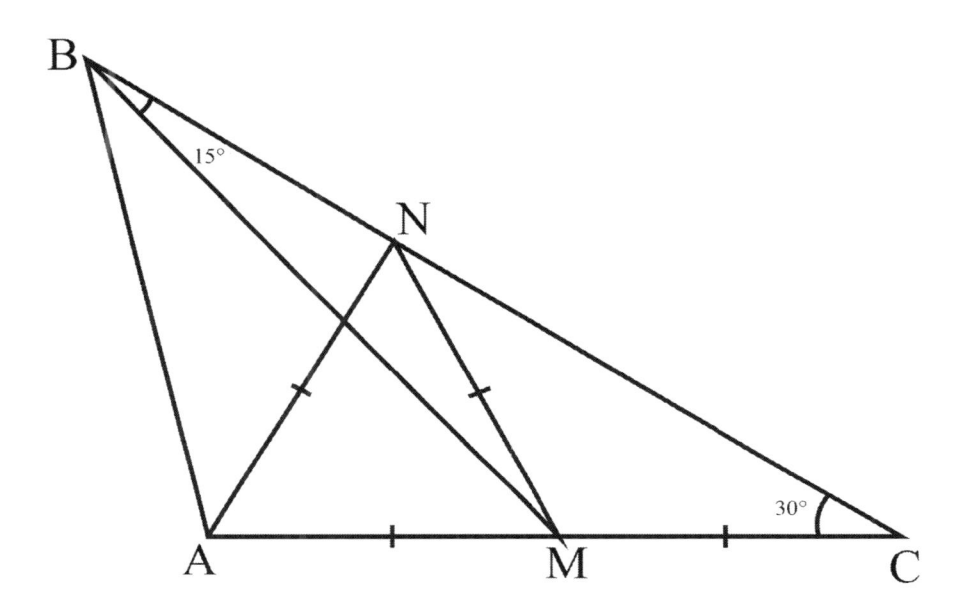

We have $\angle NAB = \angle BAC - \angle NAC = 70° - 60° = 10°$. Similarly, $\angle NCB = 10°$.

According to the first principle of the equality of triangles, $\triangle ABM = \triangle BAN = \triangle BCN$.

Thus, $\angle MAB = \angle NBA = \angle NBC = \frac{1}{2}\angle ABC = 20°$.

So, $\angle MAC = \angle BAC - \angle MAB = 70° - 20° = 50°$.

Problem 13.11: *D is a point on side BC of an isosceles triangle ABC so that BD = AC and $\angle A = \angle C = 80°$. Find $\angle BAD$.*

Answer: $10°$.

Solution: Let us build the AMC equilateral triangle.

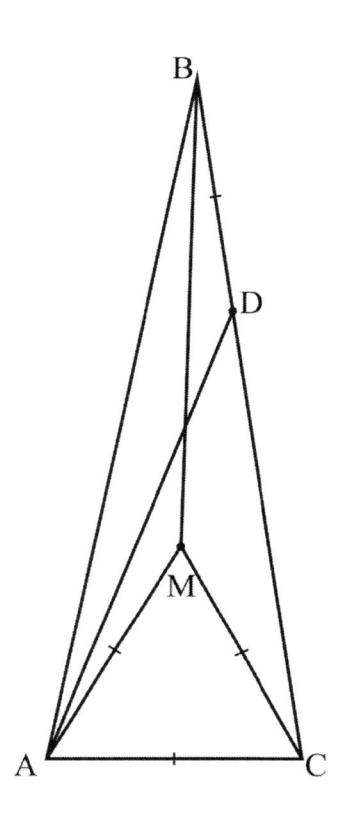

We have that $\angle MAB = \angle BAC - \angle MAC = 80° - 60° = 20°$.

Similarly, we get that $\angle MCB = 20°$.

Moreover, $\angle ABC = 180° - 2 \cdot 80° = 20°$.

Now we have that $\angle MAB = \angle DBA = \angle MCB$. According to the first principle of the equality of triangles, $\triangle DBA = \triangle MAB = \triangle MCB$.

Thus, $\angle DAB = \angle MBA = \angle MBC = \frac{1}{2}\angle ABC = 10°$.

Problem 13.12: *M is a point inside a triangle ABC with the base AC so that $\angle MBA = 10°, \angle MBC = 30°$ and $BM = AC$. Find the angles of triangle AMC. (N.M. Sedrakyan)*

> Answer: $50°, 100°, 30°$.
>
> **Solution**: Let us build the ANC equilateral triangle. Construct the MN and MC segments.
>
> From the equality of AMB and BNA triangles it follows that M and N points are in equal distances from the AB line, thus $MN \parallel AB$.
>
> From this fact it follows that
>
> $$\angle AMC = \angle NMC = \frac{360° - \angle AMN}{2} = \frac{360° - 160°}{2} = 100°.$$
>
> And $\angle MCA = 180° - (\angle MAC + \angle AMC) = 180° - (50° + 100°) = 30°$.

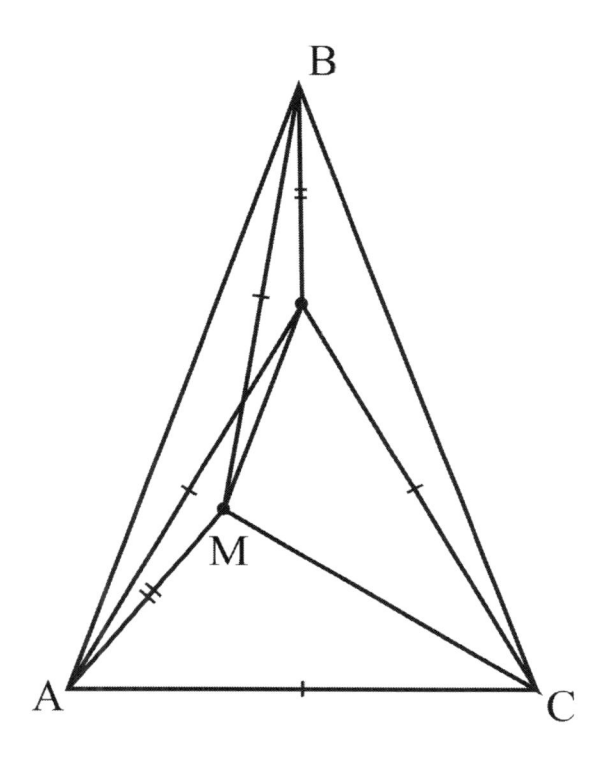

Problem 13.13: *M is a point inside an isosceles triangle ABC with the base BC so that $\angle MBC = 30°$ and $\angle MCB = 10°$. Let $\angle BAC = 80°$. Find $\angle AMC$.*

Answer: 70°.

Solution: Let us build the BNC equilateral triangle.

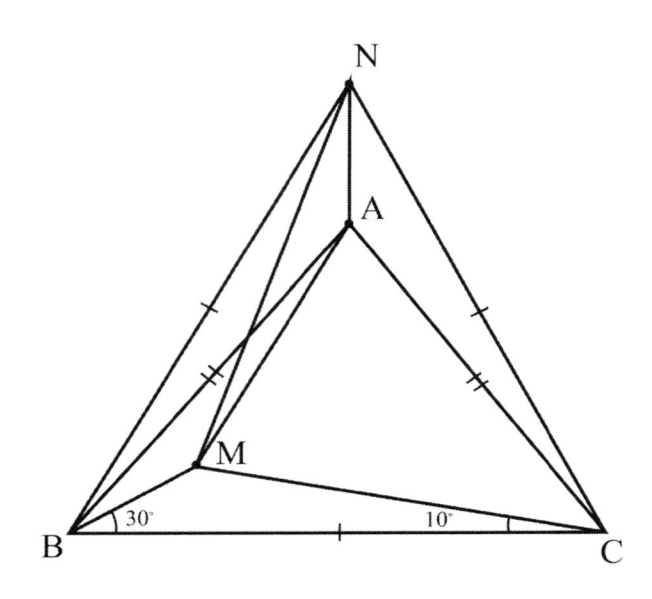

We have that $\angle ABN = \angle NBC - \angle ABC = 60° - 50° = 10°$. Similarly, $\angle ACN = 10°$.

According to the first principle of the equality of triangles, $\triangle ABN = \triangle ACN$. Thus $\angle ANB = \angle ANC = \frac{1}{2}\angle BNC = 30°$.

According to the second principle of the equality of triangles, $\triangle ABN = \triangle MCB$. Thus $AN = BM$ and $\triangle ABN = \triangle MNB$, due to the first principle of equality of triangles. So the A and M points are on equal distance from the BN line, which means that $AM \parallel BN$.

From this we have $= 180° - \angle MBN = 180° - 30° = 150°$.

$$\angle AMC = 360° - (\angle AMB + \angle BMC) = 360° - (150° + 140°) = 70°.$$

Problem 13.14: *M is a point inside a triangle ABC so that $\angle MAB = 20°$ and $\angle MBA = 10°$. Let $AB = AC$ and $\angle BAC = 100°$. Find $\angle AMC$.*

Answer: $80°$.

Solution: Let us build the BNC equilateral triangle.

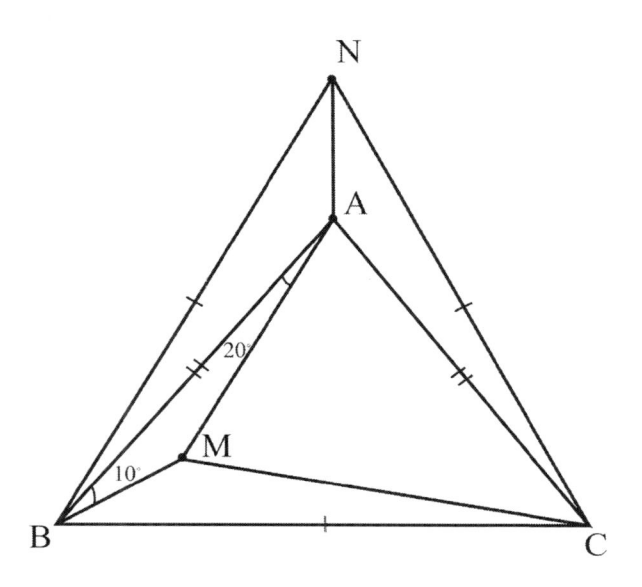

We have $\angle ABN = \angle NBC - \angle ABC = 60° - 40° = 20°$. Similarly, $\angle ACN = 20°$.

According to the first principle of the equality of triangles, $\triangle ABN = \triangle ACN$. Thus, $\angle ANB = \angle ANC = \frac{1}{2}\angle BNC = \frac{1}{2} \cdot 60° = 30°$.

We have $\angle ABN = 20° = \angle MAB$. Thus, $AM \parallel BN$.

Moreover, $\angle ANB = 30° = \angle MBA$, thus $BNAM$ is an isosceles trapezoid. From this fact it follows that $AN = BM$.

$\angle ANC = 30° = \angle MBC$, thus, according to the first principle of the equality of triangles, $\triangle MBC = ANC$, and $MC = AC$.

Then, we have $\angle AMC = \angle MAC = \angle BAC - \angle BAM = 100° - 20° = 80°$.

Problem 13.15: *ABCD is a convex quadrilateral so that $\angle BCA = 20°$, $\angle BAC = \angle DBC = 30°$ and $\angle BDC = 70°$. Prove that $AD \parallel BC$.*

Solution: Let us consider an equilateral triangle BCD (see the figure).

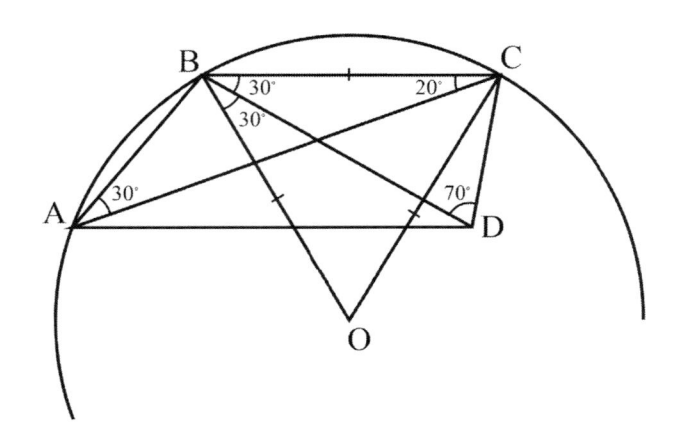

Note that point O will be the center of the circle, as $O_1B = O_1C$ and $\angle BO_1C = 2\angle BAC = 60°$.

Thus, O_1BC is an equilateral triangle. Thus, we get that $OA = OC$ and $\angle AOB = 2\angle ACB = 40°$.

We have $\angle OBD = 60° - \angle CBD = 30° = \angle DBC$. So triangles BCD and BOD are congruent triangles.

From this we get $CD = OD, \angle ODB = \angle CDB = 70°$ and $\angle BOD = \angle BCD = 180° - (30° + 70°) = 80°$.

Thus $\angle AOD = \angle AOB + \angle BOD = 40° + 80° = 120°$ and $\angle ACD = 80° - 20° = 60°$.

Note that $\angle AOD + \angle ACD = 120° + 60° = 180°$ (so, it is a cyclic quadrilateral). Thus, it follows that $\angle CAD - \angle COD = \frac{180° - \angle ODC}{2} = 20° = \angle BCA$. So $AD \parallel BC$.

Problem 13.16: *Let equilateral triangles ACM, BDN be built on the diagonals AC, BD of a convex quadrilateral ABCD, so that points B, M are on the same side of line AC, and points C, N are on the same side of line BD. Let MN = AB+CD. Find $\angle BAD + \angle CDA$. (N.M. Sedrakyan)*

> **Answer:** $120°$.
>
> **Solution:** Let us build the BCK equilateral triangle.
>
> Note that $ACB = \angle MCK$ and $\angle DBC = \angle NBK$, so, according to the first principle of the equality of triangles, $\triangle Abc = \triangle MCK$ and $\triangle DBC = \triangle NBK$.
>
> From the above mentioned fact we get that $MK = AB, \angle ABC = \angle MKC$ and $KN = CD, \angle BCD = \angle BKN$.
>
> Thus, by the problem's statement and the equations, we have $MK = AB + CD = MK + KN$. Thus, point K is on the MN segment.
>
> From the latter $\angle MKC + \angle CKN = 180°$, $\angle MKC + \angle BKN - 60° = 180°$.
>
> Thus, $240° = \angle MKC + \angle BKN = \angle ABC + \angle BCD$,
>
> and $\angle BAD + \angle CDA = 360° - (\angle ABC + \angle BCD) = 120°$.

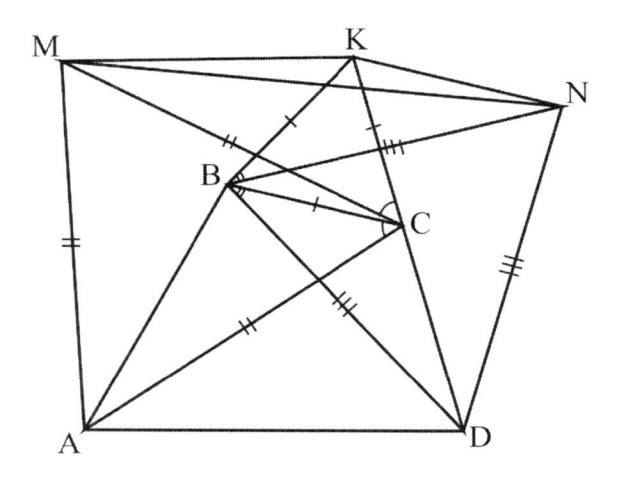

Angles and Circles

We can deduce the following properties while calculating the angles with the help of circles.

1. If AOB is a **central angle** (the vertex O coincides with the center of the circle and the sides of the angle intersect with the circle at the points A and B), then $\overset{\frown}{AB} = \angle AOB$.

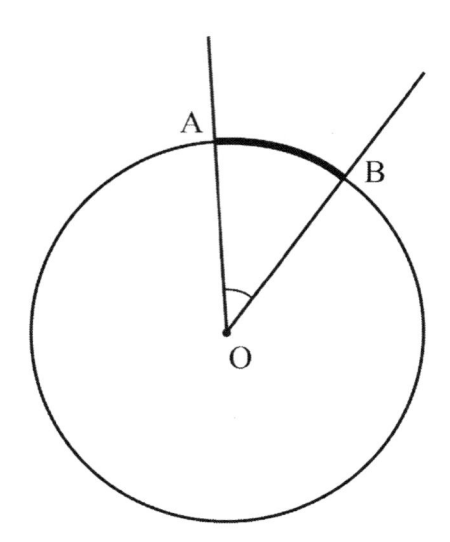

2. If ACB is an **inscribed angle** (the vertex C of the angle belongs to the circle and the points A and B intersect with the circle), then $\angle ACB = \frac{1}{2}\overset{\frown}{AB}$.

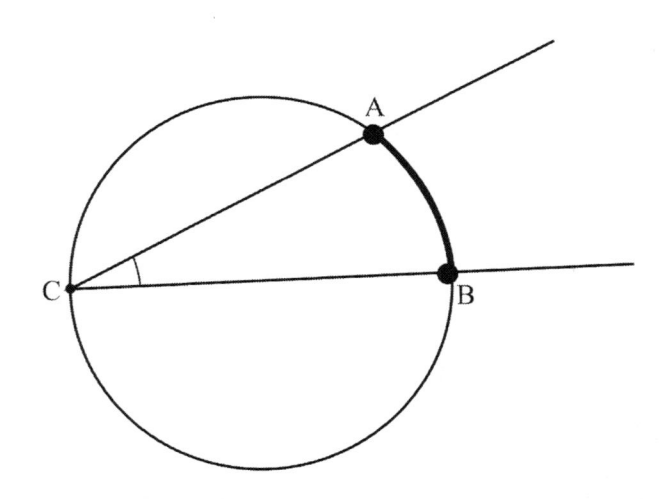

3. If chords AB and CD intersect at point M (see the figure), then $\angle AMC = \frac{1}{2}(\overset{\frown}{AC} + \overset{\frown}{BD})$.

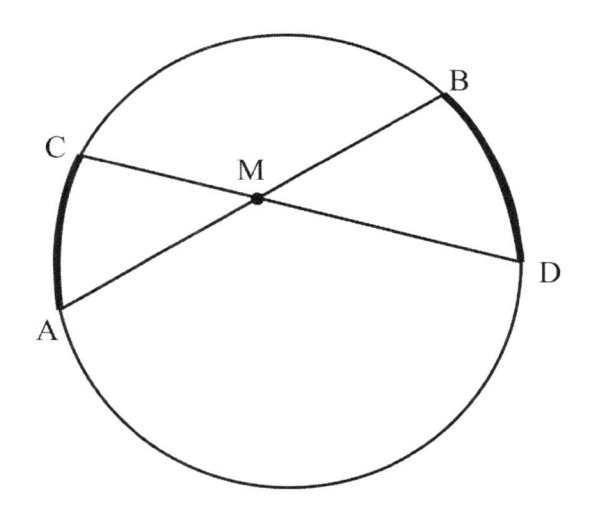

Let us look at triangle AMD.

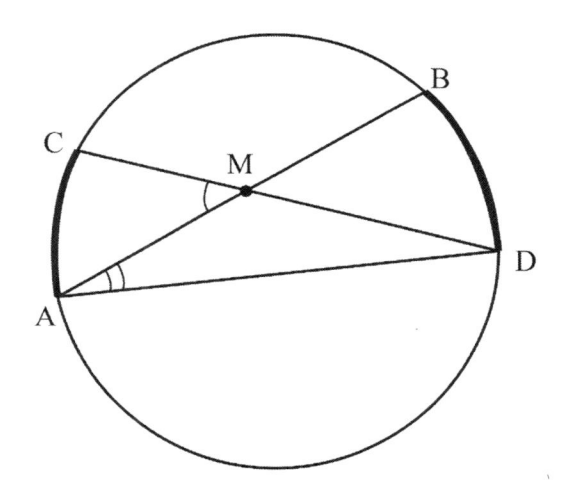

$\angle AMC$ is the outer angle of that triangle. According to the property of outer angle, $\angle AMC = \angle MAD + \angle MDA$.

We have that $\angle MAD = \angle BAD$ and $\angle MDA = \angle CDA$. According to the second property stated above, $\angle BAD = \frac{1}{2}\overset{\frown}{BD}$ and $\angle CDA = \frac{1}{2}\overset{\frown}{AC}$. Thus, $\angle AMC = \frac{1}{2}(\overset{\frown}{AC} + \overset{\frown}{BD})$.

4. Let MA and MC intersect the circle at points B and D, respectively. Then $\angle AMC = \frac{1}{2}(\overset{\frown}{AC} - \overset{\frown}{BD})$.

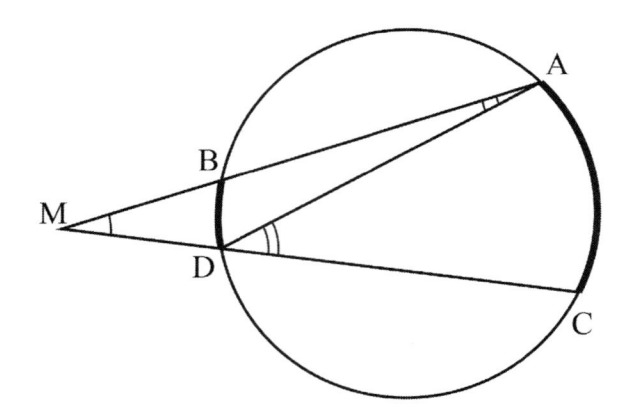

Consider triangle MAD. According to the exterior angle theorem and property 2, we have that

$$\angle ADC = \angle DAM + \angle DMA.$$

$$\frac{1}{2}\overset{\frown}{AC} = \frac{1}{2}\overset{\frown}{BD} + \angle AMC.$$

We get

$$\angle AMC = \frac{1}{2}(\overset{\frown}{AC} - \overset{\frown}{BD}).$$

5. If MA is an arc and MB is a tangent line, then $\angle AMB = \frac{1}{2}\overset{\frown}{AM}$.

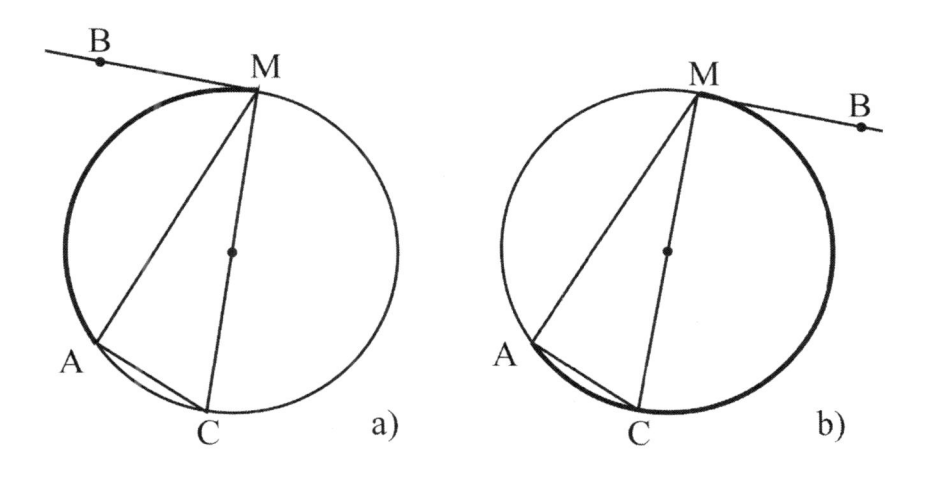

Consider diameter MC of the circle. We have that $MC \perp MB$. Thus $\angle AMB = 90° - \angle AMC$ and $\angle AMB = 90° + \angle AMC$.

We have $\overset{\frown}{MAC} = 180°$. By the second property $\angle AMC = \frac{1}{2}\overset{\frown}{AC}$.

Thus, when $\angle AMB = 90° - \angle AMC$, we have

$$\angle AMB = \frac{180° - \overset{\frown}{AC}}{2} = \frac{1}{2}\overset{\frown}{AM}.$$

Note also that

$$\angle AMB = 90° + \angle AMC = \frac{180° + \overset{\frown}{AC}}{2} = \frac{1}{2}\overset{\frown}{AM}.$$

Example 13.4:

Let O be the circumcenter of triangle ABC. Given that points O and C are on different sides of line AB line. Moreover, $\angle BOC = 80°$, $\angle AOB = 120°$. Find $\angle ABC$.

Example 13.4: Solution
Answer: $20°$.

Proof: We have $\overset{\frown}{BC} = \angle BOC = 80°$ and $\overset{\frown}{BA} = \angle AOB = 120°$.

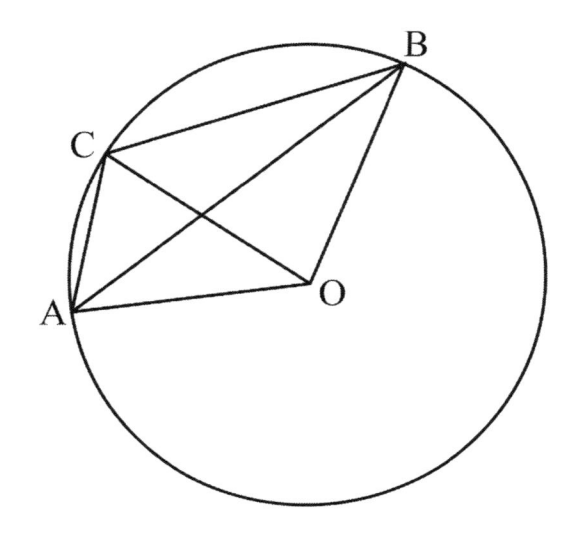

Thus, $\overset{\frown}{AC} = \overset{\frown}{BA} - \overset{\frown}{BC} = 120° - 80° = 40°$. By the second property

$$\angle ABC = \frac{1}{2}\overset{\frown}{AC} = \frac{1}{2} \cdot 40° = 20°.$$

Example 13.5:

Lemma 1. Points C and D are on the same side of line AB and $\angle ACB = \angle ADB$. Prove that points A, B, C, D are on one circle.

Example 13.5: Proof
Proof: Let us consider the following figures.

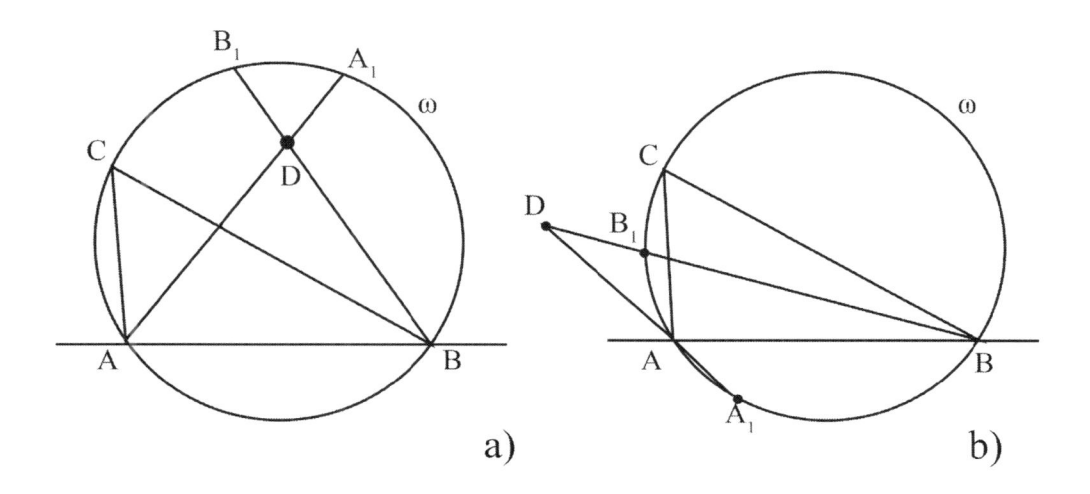

a)   b)

If we prove that point D cannot be inside or outside circle ω, then we have that point D is on the circle.

If we have the case a), then according to the third and fourth properties, we get

$$\angle ADB = \frac{1}{2}(\overset{\frown}{AB} + \overset{\frown}{A_1B_1}) > \frac{1}{2}\overset{\frown}{AB} = \angle ACB.$$

Then $\angle ADB > \angle ACB$, which is impossible.

If we have the case b), then according to the fourth and second properties, we get

$$\angle ADB = \frac{1}{2}(\overset{\frown}{A_1B} - \overset{\frown}{AB_1}) < \frac{1}{2}\overset{\frown}{A_1B} < \frac{1}{2}\overset{\frown}{AB} = \angle ACB.$$

Then $\angle ADB < \angle ACB$, which is again impossible.

13.3 Practice Problems

Problem 13.17. Let O be the circumcenter of triangle ABC and AA_1 be an altitude of triangle ABC. Prove that $\angle BAO = \angle A_1AC$.

Problem 13.18. Let O be a point on altitude AA_1 of triangle ABC. Let the circle with center O and radius OA intersects sides AB and AC at points E and F, respectively. Prove that B, E, F, C are on the same circle.

Problem 13.19. Let AB be a chord of circle ω and M be the midpoint of minor arc AB. Chords MC and MD intersect with chord AB at points E and F, respectively. Prove that E, C, F, D are on the same circle.

Problem 13.20. Let $ABCD$ be a convex quadrilateral, so that $AD = DB$ and $\angle ABC = \angle ADC = 120°, \angle ACB = 10°$. Find $\angle ADB$.

Problem 13.21. Let AA_1, BB_1, CC_1 be the altitudes of an acute triangle ABC. Let M, N, P, K be points on the sides AB, AA_1, CC_1, BC so that $B_1M \perp AB, B_1N \perp AA_1, B_1P \perp CC_1$ and $B_1K \perp BC$. Prove that M, N, P, K are on one circle.

Problem 13.22. Let circles ω_1 and ω_2 touch each other at point P, so that ω_1 does not have any points outside ω_2. Let chord AB touches circle ω_1 at point C. Prove that $\angle APC = \angle BPC$.

Problem 13.23. Given triangle ABC and circles $\omega_1, \omega_2, \omega_3, \omega_4$ passing through point A, so that centers of ω_1, ω_2 are on ray AB and centers of ω_3, ω_4 are on ray AC. Prove that four points (different from A) created by pairwise intersections of $\omega_1, \omega_2, \omega_3, \omega_4$ are on one circle (N. M. Sedrakyan).

Problem 13.24. Let $ABCD$ be a tangential quadrilateral, so that $\angle BAC = 16°, \angle DAC = 44°$ and $\angle ADC = 32°$. Find $\angle ACB$. (N. M. Sedrakyan).

13.3.1 Solutions of Practice Problems

Problem 13.17: *Let O be the circumcenter of triangle ABC and AA_1 be an altitude of triangle ABC. Prove that $\angle BAO = \angle A_1AC$.*

Solution: We have $\angle AOB = 2\angle ACB$.

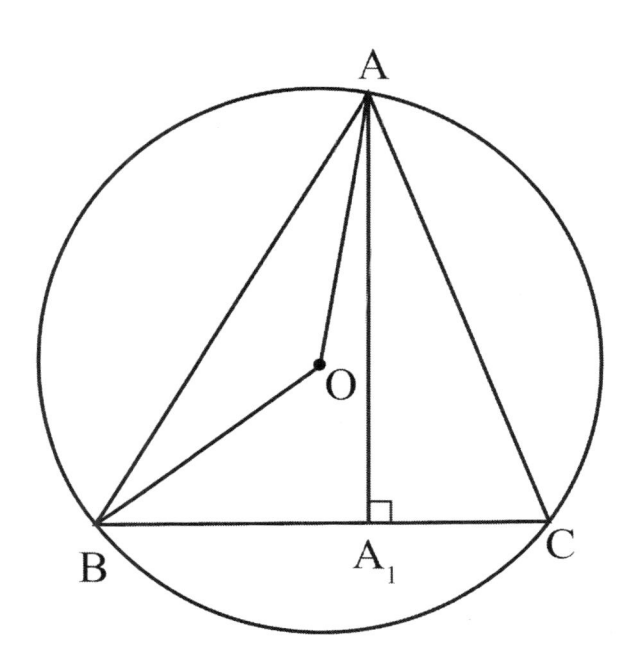

Thus, from AOB isoceles triangle we have.

$$\angle BAO = \frac{180° - \angle AOB}{2} = \frac{180° - 2\angle ACB}{2} = 90° - \angle ACA_1 = \angle A_1AC.$$

Then $\angle BAO = \angle A_1AC$.

Problem 13.18: *Let O be a point on altitude AA_1 of triangle ABC. Let the circle with center O and radius OA intersects sides AB and AC at points E and F, respectively. Prove that B, E, F, C are on the same circle.*

Solution: It is sufficient to prove that $\angle AFE = \angle EBC$.

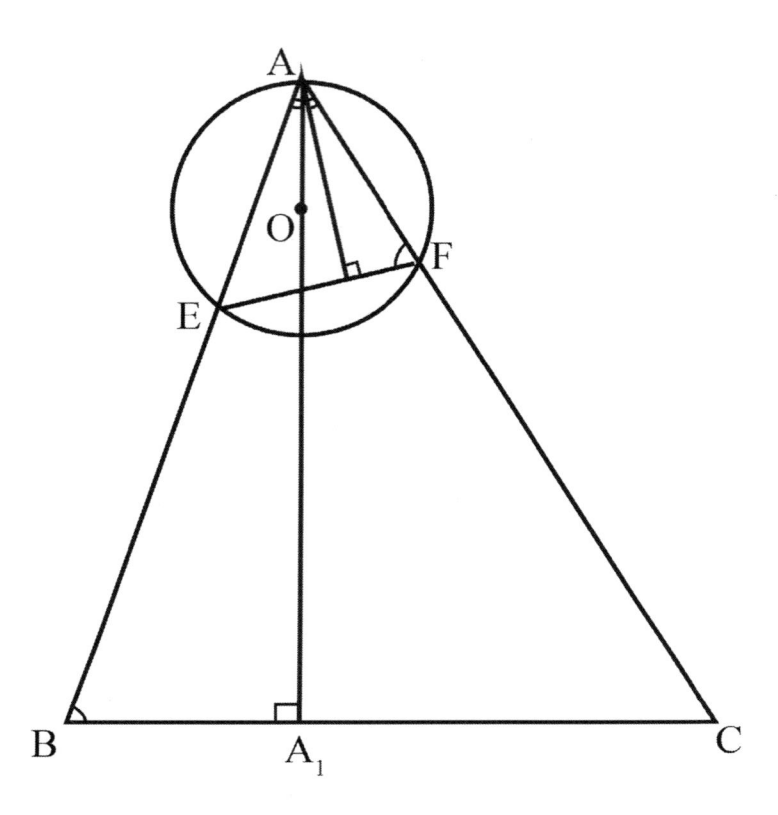

Note that the point O is the center of circle circumscribed to AEF triangle.

According to problem 13.17, we have $\angle AFE = 90° - \angle OAE = 90° - \angle A_1 AB = \angle ABC = \angle EBC$.

Problem 13.19: *Let AB be a chord of circle ω and M be the midpoint of minor arc AB. Chords MC and MD intersect with chord AB at points E and F, respectively. Prove that E, C, F, D are on the same circle.*

Solution: It is sufficient to prove that $\angle AEC = \angle CDF$.

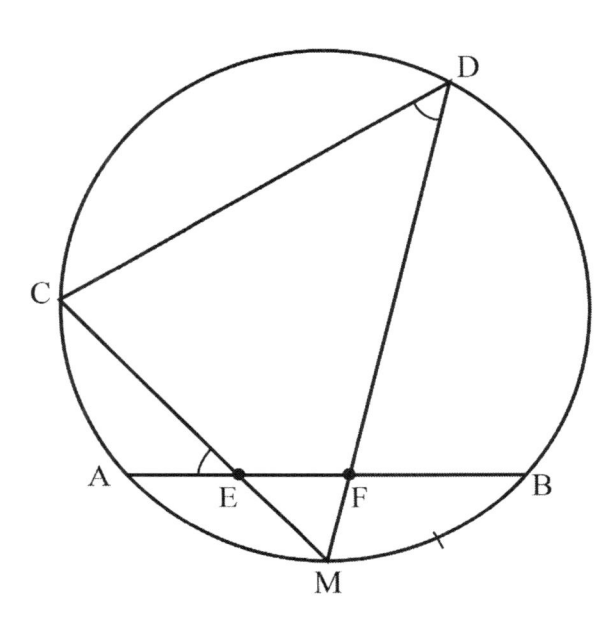

Indeed, we have $\angle AEC = \frac{1}{2}(\overset{\frown}{AC} + \overset{\frown}{BM}) = \frac{1}{2}(\overset{\frown}{AC} + \overset{\frown}{AM}) = \angle CDM = \angle CDF$.

Thus, $\angle AEC = \angle CDF$.

Problem 13.20: *Let $ABCD$ be a convex quadrilateral, so that $AD = DB$ and $\angle ABC = \angle ADC = 120°$, $\angle ACB = 10°$. Find $\angle ADB$.*

Answer: $20°$.

Solution: Consider the circumcircle of triangle ABC. Let O be its center.

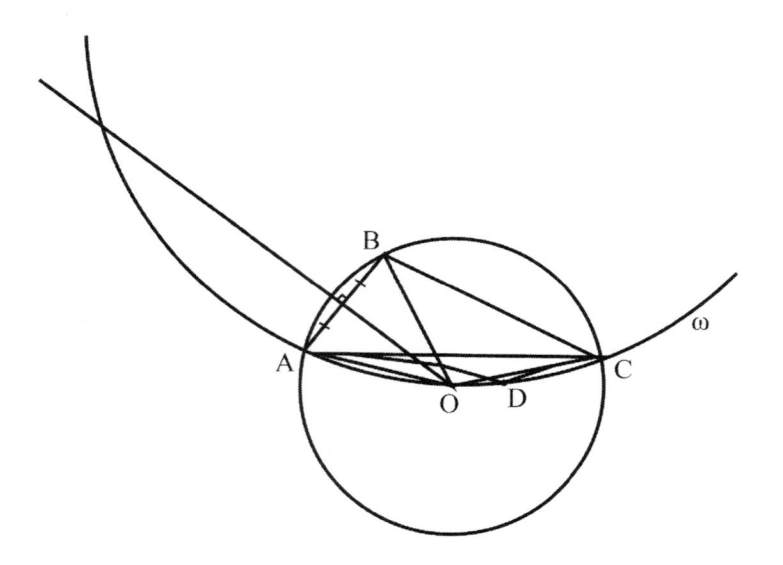

We have $\angle ABC = \frac{1}{2}\overset{\frown}{AC}$. Thus, $\overset{\frown}{AC} = 240°$.

Moreover, $\overset{\frown}{ABC} = 360° - 240° = 120°$. Thus, $\angle AOC = \overset{\frown}{ABC} = 120°$.

We know that points O and D are on the same side of line AC. $\angle AOC = 120° = \angle ADC$. According to lemma 1, points O, D, A and C are on the same circle ω.

On the other hand, O and D are on the perpendicular bisector of AB segment.

Moreover, the intersection points of perpendicular bisector of AB and circle ω should be on opposite sides of line AB.

So, points O and D coincide.

Thus, it follows that $\angle ADB = \angle AOB = 2\angle ACB = 20°$.

Problem 13.21: *Let AA_1, BB_1, CC_1 be the altitudes of acute triangle ABC. Let M, N, P, K be points on sides AB, AA_1, CC_1, BC, so that $B_1M \perp AB, B_1N \perp AA_1, B_1P \perp CC_1$ and $B_1K \perp BC$. Prove that M, N, P, K are on one circle.*

We have $\angle AMB_1 = 90° = \angle ANB_1$.

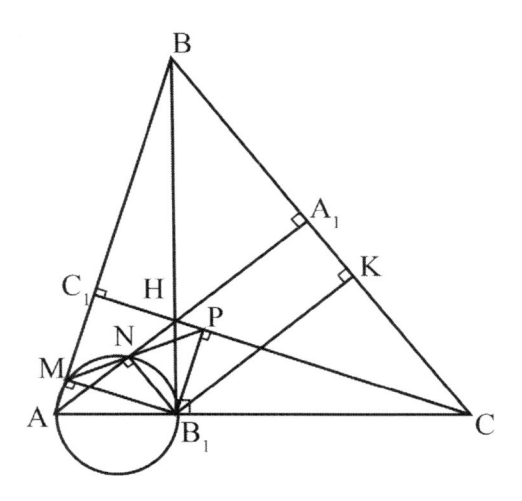

According to lemma 1, points A, M, N, B_1 are on the same circle. So

$$\angle ANM = \frac{1}{2}\widehat{AM} = \angle AB_1M$$

Note that $B_1P \perp CC_1$ and $AB \perp CC_1$, thus $B_1P \parallel AB$. Thus, it follows that $\angle MB_1P = 180° - \angle C_1MB_1 = 180° - 90° = 90°$.

We also have $\angle AB_1B = 90°$. So, $\angle AB_1B = \angle HB_1P$. Thus

$$\angle AB_1M = \angle BB_1P. \tag{13.1}$$

For quadrilateral $NHPB_1$ we have $\angle B_1NH + \angle B_1PH = 90° + 90° = 180°$. So, $NHPB_1$ is a cyclic quadrilateral. Then

$$\angle HNP = \angle HB_1P = \angle BB_1P.$$

From problems 13.1, 13.2 and 13.3 we have $\angle ANM = \angle HNP$. Thus, points M, N, P are on one line. Similarly, we prove that points K, N, P are on one line. Thus, M and K are on line NP line.

Problem 13.22: *Let circles ω_1 and ω_2 touch each other at point P, so that ω_1 does not have any points outside ω_2. Let chord AB touches circle ω_1 at point C. Prove that $\angle APC = \angle BPC$.*

Solution: Let D be the intersection point of half-line PC and circle ω_2.

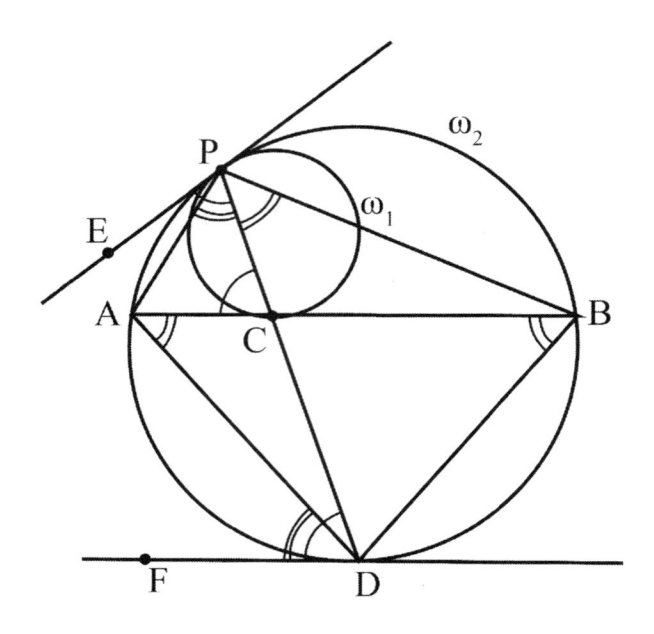

Take the PE and DF tangent lines of ω_2 circle. By fifth property we have

$$\angle ACP = \frac{1}{2}\overset{\frown}{PC} = \angle EPD = \frac{1}{2}\overset{\frown}{PD} = \angle FDP.$$

Thus, $\angle ACP = \angle FDP$, from which $AB \parallel FD$.

So, $\angle BAD = \angle ADF$. Moreover, $\angle ADF = \frac{1}{2}\overset{\frown}{AD} = \angle ABD$. Thus, $\angle BAD = \angle ABD$. By the second property

$$\angle APC = \angle APD = \frac{1}{2}\overset{\frown}{AD} =$$

$$\angle ABD = \angle BAD - \frac{1}{2}\overset{\frown}{BD} =$$

$$\angle BPD = \angle BPC.$$

Thus, $\angle APC = \angle BPC$.

Problem 13.23: *Given triangle ABC and circles $\omega_1, \omega_2, \omega_3, \omega_4$ passing through point A, so that centers of ω_1, ω_2 are on ray AB and centers of ω_3, ω_4 are on ray AC. Prove that four points (different from A) created by pairwise intersections of $\omega_1, \omega_2, \omega_3, \omega_4$ are on one circle (N. M. Sedrakyan).*

Solution: Let O_1, O_2, O_3, O_4 be the centers of $\omega_1, \omega_2, \omega_3, \omega_4$ (see the figure).

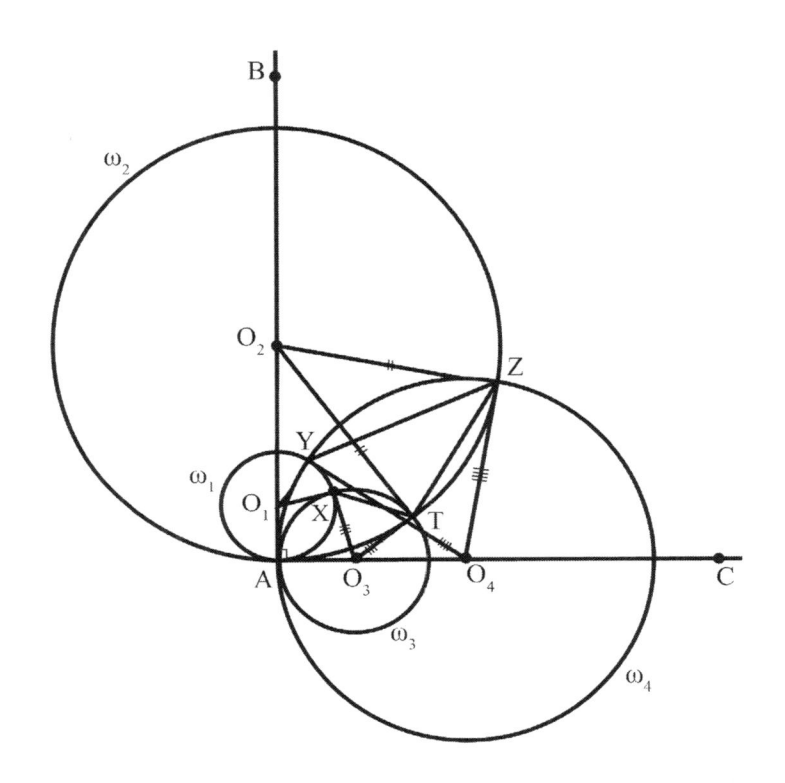

We have $O_1X = O_1Y, O_2Z = O_2T, O_3X = O_3Z, O_4Y = O_4T$ and $\angle O_1XO_2 = \angle O_1AO_3 = 90°$.
Similarly, $\angle O_1YO_4 = 90°$, $\angle O_2ZO_4 = 90°$ and $\angle O_2TO_3 = 90°$.

Suppose $\angle O_1XY = \angle O_1YX = \phi_1, \angle O_2ZT = \angle O_2TZ = \phi_2, \angle O_3XT = \angle O_3TX = \phi_3, \angle O_4ZY = \angle O_4YZ = \phi_4$.

Note that $\angle XYZ = (90° - \phi_1) + \phi_4$ and $\angle XTZ = (90° - \phi_3) + \phi_2$, $\angle YXT = 270° - \phi_1 - \phi_3$ and $\angle YZT = \phi_2 + \phi_4 - 90°$.
Then, $\angle XYZ + \angle XTZ = 180° + \phi_2 + \phi_4 - \phi_3 - \phi_1 = \angle YXT + \angle YZT$.
Thus, it follows that $\angle XYZ + \angle XTZ = 180°$.
So, X, Y, Z, T are on the same circle.

Problem 13.24: *Let ABCD be a tangential quadrilateral, so that* $\angle BAC = 16°$, $\angle DAC = 44°$ *and* $\angle ADC = 32°$. *Find* $\angle ACB$. *(N. M. Sedrakyan).*

Answer: 48°.

Solution: Let I be the incenter of quadrilateral $ABCD$ (see the figure).

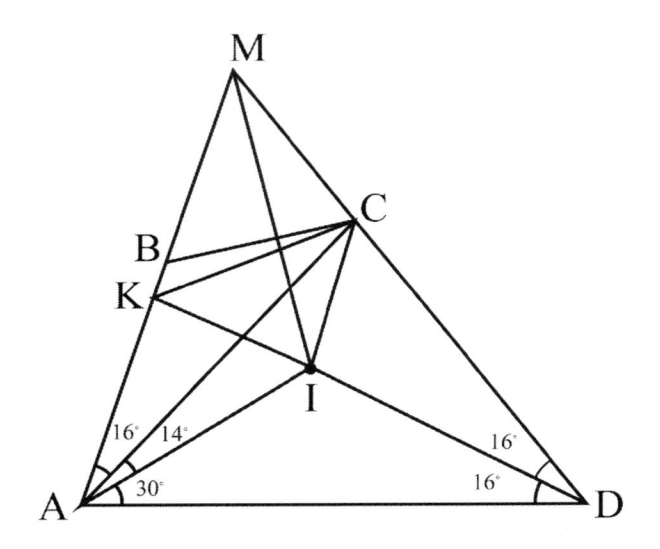

We have $\angle KAC = 16° = \angle KDC$. By lemma 1, points A, K, C, D are on one circle. So $\angle MKC = \angle ADC = 32°$.

We have $\angle MKD = \angle KAD + \angle ADK = 60° + 16° = 76°$.
So, $\angle CKD = \angle MKD - \angle MKC = 76° - 32° = 44°$. On the other hand

$$\angle CMI = \angle AMI = \frac{1}{2}\angle AMC = \frac{180° - 60° - 32°}{2} = 44°.$$

We get $\angle CKI = 44° = \angle CMI$. By lemma 1, points M, K, I, C are on one circle. Thus, $\angle ICD = \angle MKI = 76°$ and $\angle ICA = 180° - 32° - 44° - 76° = 28°$.

As $ABCD$ is a tangential quadrilateral and I is its incenter, then CI is an angle bisector and $\angle ICB = \angle ICD$. So, we get

$$\angle ACB = \angle ICB - \angle ICA = \angle ICD - \angle ICA = 76° - 28° = 48°.$$

Bibliography

[1] Sedrakyan H., Sedrakyan N., *AMC 8 preparation book*, USA (2021)

[2] Sedrakyan H., Sedrakyan N., *AMC 10 preparation book*, USA (2021)

[3] Sedrakyan H., Sedrakyan N., *AMC 12 preparation book*, USA (2021)

[4] Sedrakyan H., Sedrakyan N., *AIME preparation book*, USA (2023)

[5] Sedrakyan H., Sedrakyan N., *Number theory through exercises*, USA (2019)

[6] Sedrakyan H., Sedrakyan N., *How to prepare for math Olympiads*, USA (2019)

[7] Sedrakyan H., Sedrakyan N., *The Stair-Step Approach in Mathematics*, Springer Int. Publ., USA (2018)

[8] Sedrakyan H., Sedrakyan N., *Algebraic Inequalities*, Springer Int. Publ., USA (2018)

[9] Sedrakyan H., Sedrakyan N., *Geometric Inequalities. Methods of proving*, Springer Int. Publ., USA (2017)

[10] Spivak A., *Thousand and one math problems*, Moscow (2018)

[11] Genkin S., Itenberg I., Fomin D., *St Petersburg math circle*, (1994)

[12] Aghakhanov N., Podlipskiy O., *Math Olympiads of Moscow region*, Moscow (2006)

[13] Kurmakalin M., Kungojin A., *Zhautykov math Olympiad*, Moscow (2014)

[14] Prasolov. V., *Algebra and arithmetic problems*, Moscow (2017)

[15] Gorbachev N., *Math competition problems*, Moscow (2004)

[16] Proizvolov V., *Challenging math problems*, Moscow (2003)

Made in United States
Troutdale, OR
04/12/2024